D0622624

SPLASHDOWN!

NASA and the Navy

TURNER PUBLISHING COMPANY

Nashville, Tennessee • Paducah, Kentucky

Turner®
PUBLISHING COMPANY

412 Broadway • P.O. Box 3101
Paducah, Kentucky 42002-3101
(270) 443-0121
www.turnerpublishing.com

Turner Publishing Company Staff:
Keith R. Steele, Publishing Consultant
Steve Abell, Project Coordinator
Shelley R. Davidson, Designer

Library of Congress Control No: 2004106099

ISBN: 1-56311-985-4

Printed in the United States of America

RRH   0   9   8   7   6   5   4   3   2   1

On the cover:
Titled "The All Navy Team," this artwork is a depiction of the recovery of Apollo 12. It was painted by renowned aviation artist Stan Stokes under commission to Bob Fish.

DEDICATION

This book is respectfully dedicated to the late Charles A. "Chuck" King, Vice-President, Mutual Network, New York, who hired me in late 1965 out of Connecticut television and put me in network radio news and on five aircraft carriers for "pool" splashdown coverage.

Also to the late John Louther, a good friend who called me one day to urge that I get a demo tape to Chuck King right away. "He's hiring," said John. He was right. His phone call changed my life.

CONTENTS

Two destroyers picked up astronauts when their designated prime recovery carriers were out of position through no fault of their own. First came the destroyer Noa, doing the recovery in fine style when the carrier Randolph was too far away at the time John Glenn splashed down on Feb. 20, 1962. No debate about that and no argument over the USS Leonard Mason which raced to the Gemini 8 spacecraft carrying Neil Armstrong and Dave Scott when it made an unexpected descent into the Pacific in March of 1966 denying the recovery assignment to the carrier Boxer in the Atlantic.

Questions arise only when you look at the two other missions where destroyers became more involved than usual. First came the USS Pierce on the Scott Carpenter Mercury mission on May 24, 1962. I give the Pierce an assist on this one. Details in the section on the mission itself later on in this book.

The other mission at possible issue was Gemini 5 – astronauts Conrad and Cooper in August of 1965. Read the expanded explanation in the section on that flight. Draw your own conclusions. I call it a destroyer DuPont recovery.

SPLASHDOWN!

PRIMARILY ABOUT PRIMARY RECOVERY SHIPS

The seeds for this book were sown back in 1966 when I first stepped onto the hangar bay of the carrier USS Wasp in Boston. That was the Navy's version of a hotel check-in desk. Pictures I took then and over the next six years languished in Kodak Carousel slide trays for decades. There was an occasional slide lecture, a school here and there, a civic group breakfast, but not a whole lot more. Then, about six years ago, I began going out on cruise ships once or twice a year, freshening up my repertoire to include the Shuttle and the International Space Station, and the acceptance of those lectures re-kindled thoughts of putting my experiences down on paper.

That's what this is all about. Sixteen different recovery carriers of three different categories carrying out 31 recovery missions with the on-the-spot aid of four different destroyers when designated splashdown spots were overshot and simply could not be reached by the carriers in a safe time frame. In the mix were a chorus of helicopter squadrons and dozens of frogmen, or UDT's, from bases in Virginia and California.

Men, ships, aircraft and machines training, rehearsing and combining to safely bring back America's astronauts from Alan Shepard in his cramped little Mercury capsule to the men in the unique Apollo-Soyuz U.S. and Soviet collaboration which brought to an end what we call "The Splashdown Era." It ran from 1961 to 1975 with all astronauts recovered safely and with but one spacecraft lost to the briny deep (Grissom – Liberty Bell 7) but recovered years later.

Most of the carriers you will meet in this book were designed and built just before or during World War II. Their designations evolved as new needs arose and hulls were tweaked and decks enlarged in some cases to meet the unfolding demands of wartime or the new threats encountered in the post-war years.

By the same token, most of those noble ships have long since disappeared – most of them to the scrap yard. A few have survived and we will tell you where to find them just off the nation's coastlines. We will also direct you to every spacecraft that you will find in these pages. They all survived and NASA has done an excellent job of not putting them all under the same roof. They are spread, not only across this country, but a few have found good homes in other nations.

Other books have paid lengthy tribute to just about all the carriers you will find here and the destroyers too, for that matter. The Internet has become a treasure trove of histories of all the players unveiled on these pages. The thousands who sailed and served on those ships probably own those books or know where to find them. What we have tried to do here is bring them all together with brief summations of their great service to our country and their sacrifices. Each historical brief on a ship is followed by an equally concise description of the space mission that ship was involved in. As you will read, in a few instances, the designated prime recovery ships could not get there first through no fault of their own, but they are here nonetheless. They were ready and on station.

Our journey begins with a look at the building of this nation's space effort from decades ago and takes you through the phases that I was privileged and honored to be a part of – the manned recoveries, the practice, the rehearsals, and the human beings who made it all happen. A significant number of the photographs are my own and have never been published before. Those pictures are augmented by the often-incomparable photography of the Navy and NASA's stable of professional photographers and are thus identified.

There were fun times and good-natured incidents that I am tempted to call "Hi-jinks on the high seas," but you can judge for yourself. They always, it seemed to me, served to lighten things up and create a relaxed and cooperative atmosphere between Navy personnel of all ranks and civilians, including those of us in the news media and NASA.

To the best of our knowledge, all the above-described ingredients have never found themselves between the same covers before. We hope they provide an entertaining and informative reading experience.

WHERE DID IT ALL BEGIN?

For a moment I thought about an intro along the lines of "In the beginning...." but then quickly realized that a writer with far greater credentials than mine had already used that one.... and used it rather well.

When one thinks about rockets and rocketry and the thought spills over into fireworks and pyrotechnics, the Chinese readily come to mind, but as for real rocketry and hopes of putting people into them and into space, the folks from Peking came to that table late in the game and long after others had paved the way for all of us.

No need for an extensive history of the rocket business here. Others have come before and the territory has been expertly covered. Rare would be the fan of the space program who was not familiar with two names – Dr. Robert H. Goddard and Werner Von Braun. Both men were still in their teens when their curiosities were peaked by others before them, but their names remain at the top of the list for familiarity.

However, there was another individual whose imagination about space travel was taking flight as the 19th century was drawing to a close. He was a Russian and his name was Konstantin Tsiolkovsky. Does that name ring a bell? No surprise if it doesn't. To rehash an old cliché, Tsiolkovsky toiled in obscurity. To begin with, he had hearing problems that seemed to prevent much interaction with others – particularly in the scientific community. Instead he dreamed of space travel, of liquid-propelled rockets, but – and this is quite surprising – never built any. By the time this modest man could get anyone to agree to publish his theories in 1903 they were regarded as nothing short of madness. He would have been indeed gratified if anyone had told him.... or prophesied...that one of his countrymen, Yuri Gagarin, would become the first man in space some 58 years later.

Therefore while Tsiolkovsky was a visionary and able to conceive of machines he would never build, he did deduce that only a rocket remarkably similar to those which would, in time, take men to the Moon and back, could ever deliver the goods.

It remained for another young man living in Worcester, Massachusetts, Robert Goddard, and a few years younger than our Russian friend, to turn flights of fancy into reality. Here is where these two men separated. Goddard was the realist...the technical man, and by the time he died in 1945, he would have over two hundred rocket patents to his credit.

But even Goddard got burned by the public's knowledge of what he was up to. A paper he published in 1919 earned him the nickname "Moony," and he came to be regarded by many as just another cranky scientist with a vivid imagination. As far as anyone knows, when Goddard built and fired a liquid-propelled rocket (something about the size of a kid's toy these days) on March 16, 1926, at a farm in Auburn, Massachusetts, it marked the first ever firing of a liquid-propelled rocket. You have to remember that before Goddard lit the fuse on his handiwork, all rockets had been fueled by solid propellants.

Looking back it does not appear that Goddard could have harbored any dreams of getting rich on his labors, because when the American Interplanetary Society (now the American Rocket Society) contacted him in 1930 for information about space travel and rocketry, Goddard simply declined. Shut off from the then most renowned American source for what the Society sought, that organization looked to Europe where the German Rocket Society was moving along at a sprightly pace, building and launching their own rockets based on their own designs.

In Europe, Rumanian-born Hermann Oberth had written "The Rocket Into Interplanetary Space" in 1923, and it went on to become a classic in its genre. It also inspired another youngster, Werner Von Braun, who was just 13 when he read the book and got hooked on rockets in 1925.

As we said at the beginning of this chapter, we will not take you on a launch-for-launch march through rocket history. Suffice it to say that while Werner Von Braun would go on to build very deadly rockets for

Hitler's war machine, and his V-1 and V-2 rockets struck terror into the hearts of many British citizens beginning in the summer of 1943, Von Braun and his colleagues were smart enough to toss in the towel as the war in Europe was coming to a close. No less than 100 of them surrendered to American forces at an Inn near the Austrian border in early May of 1945. Five days later the war was over, Von Braun and his team were extensively interrogated and within months of the war's end had taken up residence at Fort Bliss, Texas, and were sowing the seeds of America's space program. Their address however, soon became the White Sands Proving Ground in New Mexico, and for the next six years, their work involved building and launching steadily improved versions of the infamous V-2.

In time even White Sands proved not large enough to encompass the sort of development Von Braun and his people were clearly capable of, and late in 1949 they opened shop in a remote and nearly uninhabited site known as Cape Canaveral in Florida.

Cycle forward just a bit to the spring of 1950 and the Army was moving its fledgling missile program from White Sands to the Redstone Arsenal near Huntsville, Alabama. The name Redstone is historically significant inasmuch as when Alan Shepard took his sub-orbital "first American in space" flight in 1961 he was rocketed into flight atop what was basically a specially modified Redstone missile.

Obviously a new space agency called NASA was reaching onto the shelves of military ordnance for the means of getting our manned space program off the launch pads at the Cape, but even as those evolutions were taking place, Von Braun and his veteran rocketeers were working on new and original monster rockets that would be needed to get this nation to the Moon and back. The winner, and-still-most-powerful rocket on Earth, would turn out to be Von Braun's greatest contribution – the mighty Saturn V.

All of a sudden the name NASA popped into this quickie look at our space program. Before anyone got real serious about manned space travel, and dreams of Moon landings and beyond began to take serious shape, our space agency, such as it was, was known as NACA – the National Advisory Committee for Aeronautics - and was established way back in 1915. Surely, its members were not just sitting around talking about lunar landings. Nonetheless, by the time the 1950's rolled around we were heavily into missile research and at least in the talking stages of space exploration. Enthusiasm for space travel was strong enough in our nation's capitol that a bill establishing the National Aeronautics and Space Administration was signed into existence on July 27, 1958, helped along doubtlessly by the USSR's audacious and frightening launch of a little humming metal ball called Sputnik on October 4, 1957, followed by Sputnik II barely a month later. It got our attention and made a lot of us very nervous. NASA opened for business on October 1, 1958.

In just a couple of years the world was introduced to the Magnificent Seven, our first group of astronauts, and the U.S. was off to the races. Even when NASA unveiled a second and third group their names remained familiar enough and appeared in our newspapers and on radio and TV often enough for excited program followers to recognize and talk about them frequently.

There you have it. A primer on the origins of this thing we call our space program. There hasn't been a "splashdown" in nearly three decades (at this writing). The last one took place in 1975. The pages that follow should be self-explanatory and are intended as a tribute to all those who waited for our astronauts, followed their returns to the Atlantic or the Pacific and brought them back safely – every time.

It was not a simple operation – far from it – but we will try to keep this explanation a lot simpler than it was in the planning and carrying out.

For a considerable amount of time before Alan Shepard became the first American in space in 1961, NASA was working on a worldwide recovery network that would allow for any and all contingencies and emergencies. It would involve ships of course, and that is where DOD (Department of Defense) played a vital role, but it also had to include a sophisticated communications network and the cooperation of other nations around the globe.

When it was finally in place, there were two main task forces – TF-140, the Atlantic Recovery Force under the control of the Atlantic Recovery Control Center at Norfolk, Virginia, and TF-130 under the control of the Pacific Recovery Control Center at Kunia, Hawaii. Suffice it to say that each of these operations was capable of moving its forces continually to maintain the best recovery positions as a spacecraft's ground track shifted during each Earth orbit.

When you look at a NASA orbital track chart like the one shown on the right it drives home a facet of the space program that most of us rarely thought about – that each time a spacecraft orbited the Earth it was moving over different territory. Mother Earth was not exactly standing still down there.

So the need was obvious from the beginning. To be able to respond to any and all conceivable landing situations including provisions for the most expeditious

Pictured at right: A section of the STS-102 Shuttle flight orbital map. (Courtesy of NASA)

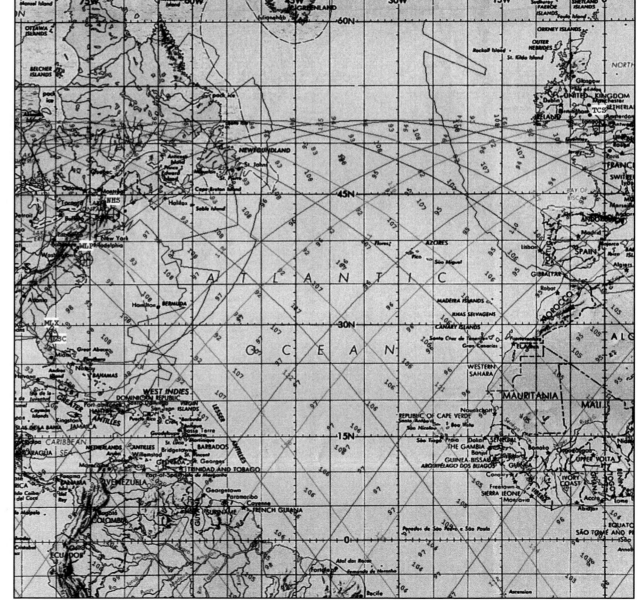

return of the spacecraft, never more clearly emphasized than during the flight of Gemini 8 when in dizzying (ask the astronauts about that one) fashion, the spacecraft had to be unexpectedly brought home, not to the Atlantic as had been originally planned but to the South Pacific. It was done and done well because NASA & the Navy were ready for it.

NASA divided these possible landing areas into five general categories and, rather than list them all, they proceeded along what you would call "degree of difficulty" parameters. One of the challenges facing NASA was spelled out by stipulations appearing in the National Space Act which said, in effect, "Make maximum utilization of the Department of Defense capabilities BUT with minimal interference with their normal operational functions." DOD would provide the ships and aircraft while NASA would come forward with the necessary special equipment. (and that would fill a couple of books).

An early appearance of so-called "special equipment" was the flotation collar that will be mentioned and shown frequently in the pages that follow. Looking back, it is hard to imagine a successful splashdown taking place without one, but the first several missions (sub-orbital) did go without and one of them (Liberty Bell 7) went to the bottom of the Atlantic. The collar made an almost seaworthy object out of what a few astronauts would call "a bad boat."

It was not always possible to come up with ground stations that could guarantee constant electronic contact with an orbiting spacecraft, although NASA came surprisingly close but did get to a point where the space agency finally had to accept communications gaps of more than the desired minimum of ten minutes. It was absolutely unavoidable and must have contributed to a few gray hairs in Mission Control.

Of course the various stations around the globe were selected on considerations of the flight plans and the Earth's rotation. DOD weighed in by allowing the use of their facilities at the Atlantic Missile Range, the Pacific Missile Range, White Sands Missile Range (New Mexico) and the Eglin Gulf Coast Test Range (Florida).

Australia got on board in allowing the use of certain existing facilities plus the construction and operation of required new facilities. In extending that courtesy, Australia put a town named Woomera on the map and generated enormous good will.

The United Kingdom permitted the construction of stations on Canton Island and Bermuda, Nigeria agreed to the lease of land and permission to construct a station in Tungu and Chawaka, and Spain didn't hurt its image as a tourist destination when it opened the door for a station in the Canary Islands.

The demands of the emerging space program were, even then, accelerating the development and improvement of a lot of voice, teletype and radar communications equipment. It is fair to say that the road to what we see today as micro-miniaturization in electronics got kick-started when NASA 's eyes turned skyward. While equipment to be used at the ground stations might have existed initially with "off the shelf" gear, the cramped quarters of Mercury, Gemini and Apollo spacecraft triggered an immediate search for smaller stuff.

The real center of this vast new worldwide system was the Computing and Communications center at the Goddard Space Flight Center in Greenbelt, Maryland. Goddard became the Master Switch for communications control, switching and distribution besides monitoring and controlling a mission from launch to landing.

Without getting into "how to build and operate your own world-wide communications center," consider that the equipment called for by this new network had to provide ground radar tracking of the spacecraft and transmission of the radar data to the computers at Goddard; launch, orbital and re-entry computations during the flight with real-time display data being transmitted to the Mercury Control Center at the Cape; real-time telemetry display data at all the sites; command capability at various stations for controlling specific spacecraft functions from the ground and, naturally, clear voice communications between the spacecraft and the ground. Big order. Big business.

Keep in mind that all these systems had to be constantly upgraded as NASA moved from Mercury to Gemini to Apollo. It was eight years from Al Shepard's sub-orbital flight to the Apollo 11 trio and many in the program, astronauts included, had to chuckle when they looked back at systems they started with and the evolution that took place by the time men walked on the Moon.

Hopefully the preceding will remind those who were there and watching the space program grow a few decades ago just how Herculean an effort was needed to get NASA up and running. The program worked; though not without difficulties and heartbreak, and it got out of the starting gate without the benefit of laptops (although an early form of them would not be long in emerging), cell phones and the various other gadgets that we just can't seem to live without today.

DOD DESCRIPTION OF THE ENTIRE RECOVERY FORCES EFFORT

To give the reader a better look at the depth of involvement in any manned mission, but especially when the program reached the APOLLO level, here is a detailed look, in the exact words of DOD (Department of Defense), on the structure and deployment of all forces under its command and control. While no two missions were identical, this material should give the reader a clear look at how many units and individuals it took to make these recoveries the overwhelming successes that they were. Not a word has been changed. This example happens to be from Apollo 15.

SUPPORT FORCES

The Department of Defense will provide approximately 2,949 persons, 38 aircraft and 4 ships, positioned around the world, for support of Apollo 15. They are from the following action and cooperating agencies:

Aerospace Rescue and Recovery Service (ARRS)
Air Force Communications Service (AFCS)
Air Force Systems Command (AFSC)
Air Force Eastern Test Range (AFETR)
Air Force Logistics Command (AFLC)
Space and Missile Test Center ((SAMTEC)
Air Weather Service (AWS)
Defense Communications Agency ((DCA)
Department of Defense Manned Space Flight Support Office (DDMS)
Military Airlift Command (MAC)
Navy Task Force 130 (TF-130)
Navy Task Force 140 (TF-140)
North American Air Defense Command (NORAD)
U.S. Coast Guard (USCG)
DOD Medical Services

The Department of Defense Manager for Manned Space Flight Support Operations, Major General David M. Jones, will exercise operational control over all supporting DOD forces at T-24 hours. The DOD Manager's support staff, under the direction of Colonel Kenneth J. Mask, will be located at the Mission Control Center and at the Alternate Control Center, Patrick Air Force Base, during operations.

NETWORK

DOD land tracking stations and the Apollo Range Instrumentation Aircraft (ARIA) join the NASA Manned Space Flight Network (MSFN) stations during Apollo missions to form a global tracking and instrumentation system. For Apollo 15, DOD network support will be provided by the Air Force Eastern Test Range (AFETR) and the North American Air Defense Command (NORAD).

The AFETR extends from the eastern United States mainland though the south Atlantic into the Indian Ocean. It includes all stations, sites, ocean areas and air space necessary to conduct missile and space vehicle testing and development. Range stations supporting the Apollo 15 mission are:

CAPE KENNEDY AIR FORCE STATION (CNV)

Data from all AFETR instrumentation sites is collected, displayed and analyzed here. Cape Kennedy will provide support in areas of C-band radar, range safety, optics, meteorology, and UHF command and destruct control.

MERRITT ISLAND (MLA)

MLA is situated just west of Cape Kennedy on the west short of the Banana River. It will provide C-band radar support from its TPQ-18 radar site for launch and Earth orbit.

GRAND BAHAMA ISLAND (GBI)

GBI, 60 miles east of Palm Beach, Florida will provide command destruct during the launch of Apollo 15.

GRAND TURK ISLAND (GTK)

Grand Turk is 90 miles north of the Dominican Republic and 660 miles southeast of Cape Kennedy. It will provide UHF command destruct during the launch phase.

ANTIGUA ISLAND (ANT)

ANT, one of the Leeward Islands, is a member of the British Commonwealth and is situated some 1,250 miles southeast of Cape Kennedy. It will provide UHF command destruct and backup VHF air/ground voice support during the launch phase.

APOLLO RANGE INSTRUMENTATION AIRCRAFT (ARIA)

ARIA are four-engined jets originally manufactured by the Boeing Company (similar to the 707 airliner/transport). They have been specifically modified by the Douglas Aircraft Company, Bendix Corporation, and the Collins Radio Company for Apollo support. The fleet of four ARIA is operated by the AFETR and the aircraft are stationed at Patrick AFB when not deployed in support of space flight.

For Apollo 15, ARIA will provide unified S-band tracking, voice relay, and both S-band and VHF telemetry during trans-lunar injection. In addition, two aircraft will be configured to send telemetry in real time via TACSAT and a MFSN SITE to the Mission Control Center for data reduction and display. ARIA will also support reentry from command and service module separation through blackout and splashdown.

Operational control of the ARIA will be maintained by the Aircraft Operations Control Center (AOCC) at Patrick AFB via the worldwide ARIA communications network.

In addition to tracking support, the AFETR also provides range safety which insures protection of life and property in the launch area should the mission go awry.

NORAD's Space Defense Center (SDC) maintains a computerized tally of man-made objects orbiting the Earth based on more than 20,000 satellite observations received daily from a network of sensors.

During Apollo 15, NORAD will use this space traffic catalog to determine Computation of Miss Distance Between Objects (COMBO). The command will then be able to furnish NASA officials information on all space objects the Apollo astronauts are likely to see, and identify all those that are actually spotted.

A COMBO program will be run through the SDC's computer prior to the mission. It will be based on a nominal mission and will provide a list of all objects that will be within 200 miles of the Earth-orbiting spacecraft. This program will be updated as needed throughout the mission.

NORAD also will provide electronic signature analysis to aid in identifying all orbiting pieces associated with the Apollo 15 launch, determining their condition, orientation, gyrations and relationship to each other. Finally, the SDC technicians will furnish Terminal Impact Prediction (TIP) for all these objects, computing when and where they will reenter the Earth's atmosphere.

—ARIA—————————————

Four EC-135N Apollo Range Instrumentation Aircraft (ARIA) will support Apollo 15 in the Pacific Ocean. They will provide coverage of the critical translunar injection burn.

ARIA will also support reentry from command and service module separation through blackout and splashdown and will receive and record telemetry from the Command Module. In addition, ARIA will provide two-way voice relay between MCC and the Apollo 15 crew during TLI and reentry.

ARIA communicates with the spacecraft through a seven-foot-diameter (dish-shaped) antenna installed in the nose of each aircraft, giving them a large, bulbous look. A worldwide network of high frequency, single sideband radio stations, land cables, submarine cables, and communications satellites provides communication between the ARIA and the MCC.

ARIA will be used for coverage of those portions of the TLI sequence from one minute before ignition through one minute after cutoff. As the TLI envelope moves with the time of day on the day chosen for the launch, the ARIA fleet must use its great mobility to "follow" this envelope. Subsequent to TLI, ARIA will remain in the Pacific to provide reentry support as mentioned above. In addition,

ARIA will provide a voice relay service in the recovery area until the astronauts leave the spacecraft. It is anticipated that this will be the TACSAT terminal on the ARIA.

The ARIA will use staging bases at Guam, Wake Island, Hawaii and elsewhere if the need arises.

C-141 jet cargo aircraft of the Military Airlift Command (MAC) will deploy from Patrick AFB, along with the ARIA, to provide maintenance support for the tracking aircraft.

—BIOASTRONAUTICS——————

Medical support personnel are selected from the Army, Navy and Air Force and represent many specialties. They are organized, trained and equipped to provide immediate and continuing medical care as required. Additionally, port liaison offices are designated at selected ports to coordinate and assist in the embarkation and debarkation of DOD medical personnel and equipment.

LAUNCH SITE MEDICAL OPERATIONS
DOD surgeons/flight surgeons and pararescue technicians will be aboard the helicopters used in the event of launch site recovery operations. The USAF Hospital at Patrick AFB will be designated as the launch site medical facility for the Apollo 15 mission and its normal staff will be augmented by highly qualified DOD medical specialists and support personnel. Aboard the M-113 rescue vehicles positioned at the launch pad will be specially trained DOD independent duty medical technicians.

FLEET RECOVERY OPERATIONS
DOD physicians and technicians, together with medical equipment, will be placed on board the recovery ships. The Primary Recovery Ship – USS OKINAWA – will have a surgical team, whereas the Second Recovery Ship – USS AUSTIN – will have an independent duty medical technician placed aboard. These individuals will render any medical treatment needed by the astronauts post recovery.

SELECTED DOD RECOVERY SUPPORT MEDICAL FACILITIES
Recovery support medical facilities are located in or near planned or contingency landing areas. These facilities are prepared to provide medical care to flight crewmembers should mission termination occur in their geographical area. The following medical facilities will maintain an alert status from launch day through completion of the Apollo 15 mission:

USAF Hospital, Patrick AFB, Florida. USAF Hospital Ramey, Ramey AFB, Puerto Rico. Medical Department, US Naval Air Station, Bermuda. USAF Hospital Lajes, Lajes Field, Azores. US Naval Hospital, San Diego, Calif. Tripler Army Medical Center, Hawaii. US Naval Hospital, Guam, Mariana Islands. US Naval Hospital Clark, Clark AB, Philippines. US Army Hospital, Okinawa, Ryukyu Islands. US Naval Hospital, Yokosuka, Japan. US Hospital Tachikawa, Tachikawa AB, Japan. 24th USAF Dispensary, Howard AFB, Canal Zone.

—RECOVERY——————

Primary responsibilities for DOD recovery forces during the Apollo 15 mission will include reentry S-band tracking, rapid location of the Command Module (CM), installation of positive flotation, flight crew recovery, and retrieval of the CM. The first priority of operations after CM location will be determining the physical condition of the crew and providing medical aid if required. The second priority will be recovery of the crew and retrieval of the CM. Both location and retrieval of the CM main parachutes and apex cover are highly desirable on a non-interference basis with higher priority operations.

LAUNCH SITE RECOVERY
The Launch Site Area is bounded by a 2 1/2 nautical mile (NM) radius circle centered on Launch Pad 39A, an eleven mile radius circle centered on the ground track thirty miles downrange, and tangential lines connecting these circles.

In the event of an abort landing in the Launch Site Area, the Launch Site Recovery Commander will direct recovery operations from one of three HH-53 helicopters stationed in the area. These helicopters, when escorted by a tanker aircraft from Eglin Air Force Base, are capable of proceeding approximately 1,000 miles downrange to conduct recovery operations.

A Utility Landing Craft (LCU), which has the capability of lifting the CM from the water and conducting shallow water salvage operations, will be stationed approximately five miles out to sea. Divers aboard the LCU have a 110-foot diving capability.

For deep-water salvage, the USS Salinan (AFT-161) will be standing by in port at the U.S. Naval Station, Mayport, Florida.

At least one of the HH-53s will carry a fire-suppression kit (FSK) suspended underneath to combat either hypergolic or brush fires. This kit is modified to carry eighty-three gallons of water in place of foam which is usually carried. Hypergolic fires cannot be extinguished until the fuels have been expended. The water in the FSK would be used to lower the temperature of a hypergolic fire so that firemen could work in the area. There are two of these modified kits available for Apollo 15.

Since this is a variable azimuth mission, recovery area will move as the launch azimuth moves from 080 to 100 degrees. Landings in this area could occur when the Launch Escape System (LES) is armed at T-42 minutes to approximately T+90 seconds ground elapsed time.

The DOD recovery forces in the Launch Site Area are:

- Three HH-53 Aerospace Rescue and Recovery Service (ARRS) helicopters attached to Detachment 15, 44 Aerospace Rescue and Recovery Squadron (ARRSq), Patrick Air Force Base, Florida.

- One LCU provided by the AFETR.

The HH-53 helicopters will assume a ground alert posture at the helicopter pads on Cape Kennedy Air Force Station when the astronauts enter the CM at T-2 hours 40 minutes. When the LES is armed at T-42 minutes, the helicopters will be airborne will be airborne in a race track pattern north of the launch pad. In addition to the flight crew, each helicopter will carry three pararescuemen and a flight surgeon to provide rescue and medical aid for the astronauts, if required.

The pararescuemen aboard one of the helicopters will be wearing Nomex suits instead of the customary wet suits. These Nomex suits are capable of withstanding flash fires of up to 2,000 degrees F and provide protection from the high temperatures of hypergolic fires and reduce hydration.

The pararescuemen will also be equipped with the Apollo swimmer radio. This is a small, compact, two channel VHF transceiver used to communicate from land or water to helicopters or astronauts.

LAUNCH ABORT RECOVERY

The Launch Abort Area is the area along the launch ground track in which landings could occur following aborts during the launch phase of the flight. The exact dimensions of this area are determined by the launch azimuth and abort performance capabilities of the spacecraft. Since the launch azimuth for Apollo 15 may vary from 80 to 100 degrees, the Launch Abort Area is bounded by a line fifty miles north of the 080 Launch azimuth ground track and fifty miles south of the 100 degree ground track. It extends from the end of the launch site area to approximately 3,400 miles downrange. This area is divided into two Sectors, A and B. Sector A extends from the end of the Launch Site Area downrange 1,000 miles and Sector B is the remainder of the area. These sectors are used to differentiate the required recovery support based on the probability of a landing in the area.

The ship and aircraft stationed in the Launch Abort Area will sweep to the south as the launch azimuth moves from 80 to 100 degrees. The recovery ship and aircraft will be positioned for optimum coverage of the 80 degree launch azimuth.

The following surface and air units will provide recovery support for the Launch Abort Area:

Surface Forces:
USS AUSTIN (LPD-4)
USS VANGUARD (T-AGM-19)

Vanguard is primarily an Apollo Instrumented Ship but may be used for recovery operations, if required. The home ports of the Austin and Vanguard are Norfolk, Virginia and Port Canaveral, Florida, respectively.

Once a successful trans-lunar injection (TLI) has occurred, the Austin will be released from recovery support.

Airborne Forces:
Station A – One HC-130
Station B – One HC-130
Station C – One HC-130

Station A will be supported by the 55th ARRSq, Eglin Air Force Base, Florida. The 54th ARRSq at Pease Air Force Base, New Hampshire, and the 57th ARRSq at Lajes Air Base, Azores, will support station B and C respectively.

PRIMARY LANDING AREA RECOVERY

The normal end-of-mission (EOM) landing area is approximately 290 miles north of Oahu, Hawaii.

The Primary Landing Area is formed by a 125 mile radius circle centered 275 miles up range from the target point, a 125 mile radius circle centered twenty-five miles up range from the target point and the tangential lines connecting the circles. The recovery force assigned to this area will be in position no later than fifteen minutes prior to the predicted CM landing time. The surface and airborne forces assigned to support the EOM recovery are listed below.

Surface Forces:

USS OKINAWA (LPH-3) will be located five miles north of the target point. In the event that an Earth orbital alternate mission is flown, the OKINAWA will also support the Pacific Zone near Hawaii. Support for the OKINAWA will be provided by the USS KAWISHIWI (AO-146).

Airborne Forces:

Specified SARAH (electronic direction-finding equipment) equipped SH-3 helicopters will carry a three-man swim team in addition to the flight crew. One swimmer on each team will be equipped with an underwater camera in addition to the regular recovery equipment. The helicopters and crews will be from Helicopter Combat Support Squadron One (HC-1), stationed at Naval Air Station, Imperial Beach, California. The swimmers are assigned to Underwater Demolition Team Eleven (UDT-11), stationed at the Naval Amphibious Base, Coronado, California.

One helicopter will be stationed ten miles up range from the target point and fifteen miles north of the CM ground track. Call Sign: SWIM 1.

One helicopter will be stationed ten miles downrange from the target point and fifteen miles north of the CM ground track. Call Sign: RECOVERY.

One helicopter will be stationed five miles south and abeam the target point. Call Sign: SWIM 2.

One helicopter will carry photographers for documentation. Call Sign: PHOTO.

RECOVERY is designated for astronaut retrieval. This helicopter will carry the personnel and the required equipment to effect the astronaut egress and retrieval.

PHOTO will be equipped with the live-TV Mini Camera and will provide the same TV coverage of recovery operations as that presented during Apollos 13 and 14.

Two HC-130 aircraft from the 76th ARRSq, Hickam AFB, Hawaii, will be stationed in the area. These aircraft will carry the ARD-17 (Cook Tracker) automatic direction-finding equipment to assist in CM location. They will also carry three pararescuemen.

One HC-130 will be positioned 165 miles up range from the target point and 100 NM abeam and to the north of the CM ground track. Call sign: HAWAII RESCUE 1.

One HC-130 will be positioned 165 miles downrange from the target point and 100 miles abeam and to the north of the CM ground track. Call sign: HAWAII RESCUE 2.

Prior to CM reentry one ARIA aircraft will be in the vicinity of the Primary Landing Area for network support. This aircraft will provide voice relay after its network requirements have been satisfied.

SECONDARY LANDING AREA RECOVERY

Two types of Secondary Landing Areas (SLA), the Earth parking orbit SLA and the deep space SLA, have been identified for the Apollo 15 mission. These landing areas will normally be located in or near one of the recovery zones (Earth orbital) or recovery lines (deep space). The SLAs are formed to include the CM target point and associated high probability dispersion area. The size of the dispersion area will depend on whether the CM is entering from a parking orbit or deep space.

Earth Parking Orbit Phase

Earth Parking Orbit (EPO) Second Landing Areas are for low-speed entries from the parking orbit. The SLA is a 210 NM-long by 80 NM-wide ellipse with the major axis coincident with the entry ground track and centered on the target point. For the Apollo 15 mission, parking orbit SLAs will be located in the Atlantic and Pacific recovery zones.

Deep Space Phase

Deep space SLAs are designed for high-speed entries from deep space. They are defined for the trans-lunar, lunar orbit and trans-Earth phases of the mission and are positioned along ship-supported recovery lines. Deep space SLA's dimensions are identical to the EOM area dimensions. For the Apollo 15 mission, deep space SLAs will normally be targeted to a pre-designated area west of Hawaii. The OKINAWA will be on station in the area to provide support. Deep space aborts to other areas will be considered contingency conditions.

Contingency Landing Area Recovery

The Contingency Landing Area for the Earth parking orbital phase include the Earth's surface between 34 degree N and 34 degrees S latitude, except for the previously specified recovery areas. The deep space contingency landing area includes the Earth's surface between 40 degrees N and 40 degrees S latitude. Forces required to support the contingency landing area for the Apollo 15 mission are the Aerospace Rescue and Recovery Surface SAR alert aircraft which are stationed at strategic bases around the world.

Post-landing Activity

Apollo 15 recovery operations will not include the quarantine precautions that were taken during the previous lunar mission recovery operations. Once the swimmers have inflated the flotation collar and attached the life raft, the astronauts will transfer to the life raft and be picked up by the recovery helicopter. Once aboard the helicopter, the astronauts will be flown to the OKINAWA where they will undergo a quick but comprehensive medical examination. During the medical examination, the OKINAWA will retrieve the CM and then proceed at best speed to Pearl Harbor. Six hours after recovery or the following morning, depending on the decision of the medical team, the astronauts will be flown by helicopter to Hickam Air Force Base, Hawaii, where they will board a C-141 aircraft and be flown to Ellington Air Force Base, Texas, for tranfer to the NASA MSC.

Approximately four hours after recovery the first lunar sample shipment will be flown by helicopter from the PRS to Hawaii. From Hawaii, an ARIA will fly the shipment to Ellington Air Force Base, Texas. The second lunar shipment will accompany the astronauts from Hawaii to Ellington.

—WEATHER

More than 200 DOD meteorologists and space environment specialists are deployed to forward bases and are assigned to direct in-place support of Apollo 15. DOD weathermen around the world are observing environmental conditions at the Earth's surface, in the upper atmosphere, and in space itself. This information is evaluated and quickly relayed to the Mission Control Center in Houston for use by NASA officials directing the flight.

The U.S. Air Force Air Weather Service (AWS) will employ its extensive meteorological network along the Air Force Eastern Test Range, southward from Cape Kennedy to Ascension Island, to provide conventional weather support for the Apollo 15 launch. In addition to normal hourly surface observation, many critical soundings of the upper atmosphere will be made during the last two days of the launch countdown.

Beginning three days before splashdown, the AWS 9th Weather Reconnaissance Wing will undertake vital weather reconnaissance of the primary recovery zone using WC-135 weather aircraft staging from American Samoa. On two of the previous manned Apollo missions, this recovery area reconnaissance provided information that resulted in a late hour shifting of the splashdown point to an area with safe weather conditions. The 9th Weather Reconnaissance Wing will also provide WC-130 and WC-135 weather reconnaissance over the launch site at Cape Kennedy with particular emphasis on detection of clouds with potential for electrical discharges.

The capability of one of the largest meteorological computers in the free world at the Air Force Global Weather Central, Offutt Air Force Base, Nebraska, will be accessible to DOD meteorologists for use in Apollo support.

The AWS support capability also extends into outer space. The AWS Space Environmental Support System (SESS), backbone of which is a global chain of seven strategically located optical and radio telescopes and sensing devices, gathers data on solar radiation and energetic particle emission which may adversely affect manned operations in space.

This information also bears on the effective management of communications networks as various types of intense solar activity can cause high frequency radio blackouts. Focal point of SESS is the AWS Aerospace Environmental Support Center, (AESC), Cheyenne Mountain, Colorado, which routes significant solar data to Houston via the National Oceanic and Atmospheric Administration's Space Environmental Services Center, Boulder, Colorado.

An AWS detachment at Sunnyvale, California, provides additional specialized space environmental support involving energetic particle, X-ray and solar wind information obtained from radiation sensing satellites. Through this round-the-clock watch on the space environment, AWS weathermen are able to provide NASA with the vital advance warnings necessary to compensate for potentially disruptive solar activity.

The U.S. Navy's recovery task force will furnish Houston with special Apollo weather reports depicting sea and weather conditions in the primary or alternate recovery zones. The Navy facility at Barking Sands, Hawaii, will support scientific studies by launching a Loke Dartsonde meteorological rocket at the time of the Command Module reentry into the Earth's atmosphere.

So ends this verbatim look at most of the DOD's Apollo 15 guidebook. All we didn't reproduce here were biographies and detail on the ships and planes most closely involved. As I typed it out I was impressed and exhausted just contemplating the depth of involvement by all those units fanned out across the globe. At the very beginning of this section DOD noted just four (4) ships involved. As our accompanying listing of all the ships involved in all the missions also illustrates, the fleet sizes were rapidly boiling down to just a handful of ships as the splashdown world was coming to a close.

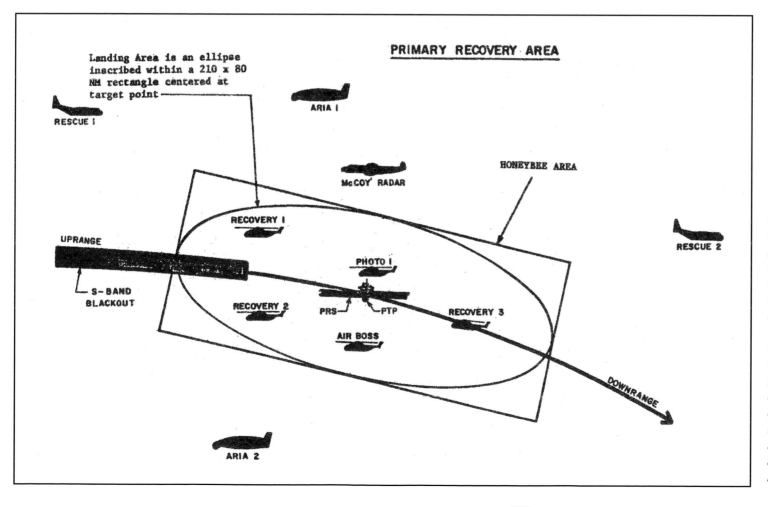

Landing Area is an ellipse inscribed within a 210 x 80 NM rectangle centered at target point

PRIMARY RECOVERY AREA

RESCUE 1

ARIA 1

McCOY RADAR

HONEYBEE AREA

RECOVERY 1

RESCUE 2

UPRANGE

S-BAND BLACKOUT

PHOTO 1

RECOVERY 2

PRS — PTP

RECOVERY 3

AIR BOSS

DOWNRANGE

ARIA 2

Purely for the purpose of example here is a visual chart of a typical Apollo mission Primary Recovery Area. As you can see, helicopters are doing the close-in work and there are five of them deployed. Air Boss, which had been fixed-wing on most of the Atlantic recoveries, operated much like a film director. It tied the operation together and tended to be a helicopter in missions where the splashdown point was far from a land area.

Rescue 1 and 2 would be farther out on recovery day and Air Force Para rescue teams would be on call, and ready to be used, as needed. These aircraft would also be carrying automatic direction-finding equipment and would be an integral part of the whole airborne recovery force. ARIA 1 and 2 were basically communications relay aircraft between the Command Module and Mission Control. (Courtesy of NASA)

SHIPS USED IN THE SPACE PROGRAM

MERCURY-ATLAS 1 (Unmanned)
JULY 29, 1960

Atlantic:
USS Hailey – DD-556. Destroyer
USS Power – DD-839. Destroyer
USS Vesole – DD-878. Destroyer
USS Manley – DD-940. Destroyer
USS Robert H. McCard – DD-822. Destroyer
USS Casa Grande – LSD-13. Landing ship dock
USS Escape – ARS-6. Salvage ship

MERCURY-REDSTONE 1-A (Unmanned)
DEC. 19, 1960

Atlantic:
USS Valley Forge – LPH-8. Prime recovery ship
USS Conway – DD-507. Destroyer
USS Cony – DD-508. Destroyer
USS Waller – DD-466. Destroyer
USS Eaton – DD- 510. Destroyer
USS Bache – DD-470. Destroyer
USS Beale – DD-471. Destroyer
USS Murray – DD-576. Destroyer
USS Perry – DD-844. Destroyer
USS Escape – ARS-6. Salvage ship

MERCURY-ATLAS 2 (Unmanned)
FEB. 21, 1961

Atlantic:
USS Donner – LSD-20. Prime recovery ship
USS Eugene A. Greene – DD-711. Destroyer

USS Bordelon – DD-881. Destroyer
USS Joseph P. Kennedy – DD-850. Destroyer
USS Borie – DD-704. Destroyer
USS Opportune – ARS-41. Salvage ship

MERCURY-REDSTONE 2 (Crew: Ham, the chimpanzee)
JAN. 31, 1961

Atlantic:
USS Donner – LSD-20. Prime recovery ship
USS Manley – DD-940. Destroyer
USS Harold J. Ellison – DD-864. Destroyer
USS Cone – DD-866. Destroyer
USS Robert H. McCard – DD-822. Destroyer
USS Warrington – DD-843. Destroyer
USS Borie – DD-704. Destroyer
USS Opportune – ARS-41. Salvage ship

MERCURY-ATLAS 3 (Unmanned)
APRIL 25, 1961

Destroyed on lift off.
Atlantic:
USS Eugene A. Greene – DD-711. Destroyer
USS Gyatt – DD-712. Destroyer
USS Furse – DDR-882. Destroyer
USS Basilone – DD-824. Destroyer
USS Steinaker – DD-863. Destroyer
USS Joseph P. Kennedy – DD-850. Destroyer
USS Purdy – DD-734. Destroyer
USS Hyman – DD-732. Destroyer
USS Beatty – DD-756. Destroyer

USS Bristol – DD-857. Destroyer
USS Donner – LSD-20. Landing ship dock
USS Chukawan – AO-100. Oiler
USS Recovery – ARS-43. Salvage ship

MERCURY-REDSTONE 3, FREEDOM 7 – ALAN SHEPARD
MAY 5, 1961

Since this mission was sub-orbital and measured in minutes there was no need for a Pacific deployment. All ships listed here, therefore, were out in the Atlantic in areas of probable splashdown.

USS Lake Champlain – CVS-39. Prime recovery carrier.
USS Abbot – DD-629. Destroyer
USS Ability – MSO-519. Minesweeper
USS Decatur – DDG-31. Guided missile destroyer
USS Newman K. Perry – DD-883. Destroyer
USS Notable – AM-267. Minesweeper
USS Recovery – ARS-43. Salvage ship
USS Rooks – DD-804. Destroyer
USS Sullivans – DD-537. Guided missile destroyer – Named in honor of the five brothers who all perished on the same ship in WWII.
USS Wadleigh – DD-689. Destroyer

MERCURY-REDSTONE 4, LIBERTY BELL 7 – GUS GRISSOM
JULY 21, 1961

The second and last sub-orbital. Again, all ships were dispersed in the Atlantic.

USS Randolph – CVS-15. Prime recovery carrier
USS Alacrity – MSO-520. Minesweeper ocean
USS Conway – DD-507. Destroyer
USS Cony – DD-508. Anti-sub destroyer
USS Exploit – MSO-440. Minesweeper ocean
USS Lowry – DD-770. Destroyer
USS Recovery – ARS-43. Salvage ship
USS Stormes – DD-780. Destroyer

MERCURY-ATLAS 4 (Unmanned)
SEPT. 13, 1961

Atlantic:
USS Plymouth Rock – LSD-29. Prime recovery carrier
USS Decatur – DDG-31. ACTUAL recovery ship
USS Glennon – DD-840. Destroyer
USS Wallace L. Lind – DD-703. Destroyer
USS Borie – DD-704. Destroyer
USS Cony – DD-508. Anti-sub destroyer
USS Sarsfield - DD-837. Destroyer
USS Zellars – DD-777. Destroyer
USS Bigelow – DD-942. Destroyer
USS Escape – ARS-6. Salvage ship

MERCURY-ATLAS 5 (Crew: Enos, the chimpanzee)
NOV. 19, 1961

Atlantic:
USS Lake Champlain – CVS-39. Prime recovery carrier
USS Stormes – DD-780. ACTUAL recovery ship
USS Cone – DD-866. Destroyer
USS Chikaskia – AO-54. Oiler
USS Fiske – DD-842. Destroyer
USS Lowry – DD-770. Destroyer
USS Laffey – DD-724. Destroyer
USS Hawkins – DD-873. Destroyer
USS Witek – EDD-848. Destroyer
USS Perry – DD-844. Destroyer
USS Compton – DD-705. Destroyer
USS Bigelow – DD-942. Destroyer
USS Blandy – DD-943. Destroyer
USS Vogelgesang – DD-862. Destroyer
USS John Willis – DE-1027. Destroyer
USS Fort Mandan – LSD-21. Landing ship dock
USS Hoist – ARS-40. Salvage ship
USS Fidelity – MSO-443. Minesweeper ocean

MERCURY-ATLAS 6, FRIENDSHIP 7 – JOHN GLENN
FEB. 20, 1962

The first American in orbit, so for the first time there was a ship in the Pacific. This and the two missions that followed represented the high water mark in the number of ships placed in potential splashdown paths.

Atlantic:

USS Noa – DD-841. Destroyer, but the ACTUAL recovery ship
USS Randolph – CVS-15. Prime recovery carrier
USS Bearss – DD-654. Destroyer
USS Barry – DD-933. Destroyer
USS Blandy – DD-943. Destroyer
USS Brownson – DD-868. Destroyer
USS Chuckawan – AO-100. Oiler
USS Cone – DD-866. Destroyer
USS Exploit – MSO-440. Minesweeper
USS Glennon – DD-840. Destroyer
USS Goodrich – DDR-831. Radar picket destroyer
USS Hugh Purvis – DD-709. Destroyer
USS Kenneth D. Bailey – DD-713. Destroyer
USS Norfolk – DL-1. Hunter killer frigate
USS Observer – AM-461. Minesweeper
USS Recovery – ARS-43. Salvage ship
USS Sarsfield – DDE-837. Anti-sub destroyer
USS Charles S. Sperry – DD-697. Destroyer
USS Stormes – DD-780. Destroyer
USS Stribling – DD-876. Destroyer
USS Turner – DD-834. Destroyer
USS Witek – EDD-848. Destroyer
USS Forrestal – CVA-59. Carrier

Pacific:

USS Antietam – CVS-36. Carrier

MERCURY-ATLAS 7, AURORA 7 – SCOTT CARPENTER
MAY 24, 1962

Atlantic:

USS John R. Pierce – DD-753. Destroyer, and the ACTUAL recovery ship

USS Intrepid – CVS-11. Prime recovery carrier
USS Barton – DD-722. Destroyer
USS Dewey – DLG-14. Guided missile frigate
USS Donner – LSD-20. Landing ship dock
USS Elokomin – AO-55. Oiler
USS English – DD-696. Destroyer
USS Farragut – DLG-6. Destroyer
USS Forrestal – CVA-59. Carrier*
USS Fred T. Berry – DDE-858. Anti-sub destroyer
USS Hank – DD-702. Destroyer
USS Hoist – ARS-40. Salvage ship
USS Massey – DD-778. Destroyer
USS Moale – DD-693. Destroyer
USS Robinson – DD-562. Destroyer
USS Soley – DD-707. Destroyer
USS Spiegel Grove – LSD-32. Landing ship dock
USS Sturdy – MSO-494. Minesweeper ocean
USS Swerve – MSO-495. Minesweeper ocean – Makes you wonder where they got some of these names…and why.

Pacific:

USS Remey – DD-688. Destroyer
USS Wren – DD-568. Destroyer
USS Hunt – DD-674. Destroyer

*Forrestal appears on a list supplied by NASA's Washington, D.C. History Office for this mission, but not in the Apollo 17 Press Kit from which most of this list was compiled. We left it in.

MERCURY-ATLAS 8, SIGMA 7 – WALLY SCHIRRA
OCT. 3, 1962

Atlantic:

USS Affray – MSO-511. Minesweeper ocean
USS Alacrity – MSO-520. Minesweeper ocean
USS Barry – DD-933. Destroyer
USS Bordelon – DDR-881. Radar picket destroyer
USS Charles Adams – DDG-2. Guided missile destroyer
USS Charles S. Sperry – DD-697. Destroyer
USS Decatur – DDG-31. Guided missile destroyer

USS Dyess – DDR-880. Radar picket destroyer
USS Fletcher – DD-445. Destroyer
USS Fred T. Berry – DDE-858. Anti-sub destroyer
USS Furse – DDR-882. Radar picket destroyer
USS Haynesworth – DD-700. Destroyer
USS Lake Champlain – CVS-39. Carrier
USS Henley – DD-391. Destroyer
USS Hoist – ARS-40. Salvage ship
USS Independence – CV-62. Carrier
USS Ingraham – DD-694. Destroyer
USS John Paul Jones – DD-32. Destroyer
USS Kaskaskia – AO-27. Oiler
USS Norris – DDE-859. Anti-sub destroyer
USS Willard Keith – DD-775. Destroyer

Pacific:
USS Kearsarge – CVS-33. Prime recovery carrier
USS Epperson – DDE-719. Anti-sub destroyer
USS O'Bannon – DDE 450. Destroyer
USS Phillip – DD-498. Destroyer
USS Radford – DDE-446. Anti-sub destroyer
USS Walker – DDE-517. Anti-sub destroyer

MERCURY-ATLAS 9, FAITH 7 – L. GORDON COOPER
MAY 15-16, 1963

Atlantic:
USS Adroit – MSO-509. Minesweeper
USS Beatty – DD-756. Destroyer
USS Compton – DD-705. Destroyer
USS Davis – DD-937. Destroyer
USS Gainard – DD-706. Destroyer
USS Harwood – DD-861. Destroyer
USS Hyman – DD-732. Destroyer
USS Myles C. Fox – DD-829. Destroyer
USS Opportune – ARS-41. Salvage ship
USS Stalwart – AM-493. Minesweeper
USS Wasp – CVS-18. Carrier – A back up on this mission. Wasp
   would go on to be prime recovery carrier for 5 Gemini missions
   as you will see.

Pacific:
USS Kearsarge – CVS-33. Prime recovery carrier
USS DeHaven – DD-727. Destroyer
USS Duncan – DDR-874. Radar picket destroyer
USS Epperson – DD-719. Destroyer
USS Fletcher – DD-445. Destroyer
USS Frank Knox – DD-742. Destroyer – Knox was FDR's Navy secretary.
USS Chipola – AO-63. Oiler
USS John A. Bole – DD-755. Destroyer
USS John. W. Thomason – DD-760. Destroyer
USS Kawishiwi – AO-146. Oiler
USS Lofberg – DD-759. Destroyer
USS Mansfield – DD-728. Destroyer
USS Taussig – DD-746. Destroyer

*Twenty-four ships for this mission— and 14 of them were destroyers
and one was a radar picket destroyer. Popular ships.*

Now we move on to the Gemini two-man flights.

GEMINI-TITAN 1 – None
APRIL 8, 1964

GEMINI-TITAN 2 (Unmanned)
JAN. 19, 1965

Atlantic:
USS Lake Champlain – CVS-39. Prime recovery carrier
USS Agile – MSO-421. Minesweeper ocean
USS Bulwark – MSO-425. Minesweeper ocean
USS Paiute – ATF-159. Fleet ocean tug
USS O'Hare – DD-889. Destroyer
USS Holder – DD-819. Destroyer
USS Vogelgesang – DD-862. Destroyer
USS Putnam – DD-757. Destroyer
USS Forrest Royal – DD-872. Destroyer
USS Eugene A. Greene – DD-711. Destroyer

GEMINI-TITAN 3, MOLLY BROWN – GUS GRISSOM AND JOHN YOUNG
MARCH 23, 1965

Atlantic:

USS Intrepid – CVS-11. Prime recovery carrier

USS Bigelow – DD-942. Destroyer

USS Boston – CA-69. Heavy cruiser

USS Cony – DD-508. Anti-sub destroyer

USS Diligence – WMEC-616. Cutter

USS Douglas H. Fox – DD-779. Destroyer

USS Harold J. Ellison – DD-864. Destroyer

USS Harwood – DD-861. Destroyer

USS John Paul Jones – DD-32. Guided missile destroyer

USS Kankakee – AO-39. Oiler

USS Mullinix – DD-944. Destroyer

USS Nipmuc – AFT-157. Fleet ocean tug

USS Rich – DD-820. Destroyer

USS Robert L. Wilson – DD-847. Destroyer

USS Robert Owens – DD-827. Destroyer

USS Sarsfield – DDE-837. Anti-sub destroyer

USS Sturdy – MSO-494. Minesweeper ocean

USS Swerve – MSO-495. Minesweeper ocean

USS Vigilant – WPC-617. Gunboat

USS Ault – DD-698. Destroyer

GEMINI-TITAN 4 – JIM McDIVITT AND ED WHITE
JUNE 3 - 7, 1965

Atlantic:

USS Wasp – CVS-18. Prime recovery carrier – The first of five such missions.

USS Barry – DD-993. Destroyer

USS Blandy – DD-943. Destroyer

USS Charles S. Sperry – DD-697. Destroyer

USS Chukawan – AO-100. Oiler

USS Furse – DD-882. Destroyer

USS Hawkins – DD-873. Destroyer

USS Hoist – ARS-40. Salvage ship

USS Nimble – MSO-459. Minesweeper

USS Rich – DD-820. Destroyer

USS Robert A. Owens – DD-827. Destroyer

USS Skill – MSO-471. Minesweeper ocean

Pacific:

USS Orleck – DD-886. Destroyer

USS Ponchatoula – AO-148. Oiler

USS Rupertus – DD-851. Destroyer

USS Leonard F. Mason – DD-852. Destroyer – Destined for later fame in the program.

USS Higbee – DD-806. Destroyer

USS Goldsborough – DDG-20. Guided missile destroyer

GEMINI-TITAN 5 – L. GORDON COOPER AND CHARLES 'PETE' CONRAD
AUG. 21 - 29, 1965

Atlantic:

USS Lake Champlain – CVS-39. Prime recovery carrier

USS DuPont – DD-941. Destroyer, whose divers did the actual recovery

USS Avenge – MSO-423. Minesweeper ocean

USS Exultant – MSO-441. Minesweeper ocean

USS James C. Owens – DD-776. Destroyer

USS John W. Weeks – DD-701. Destroyer

USS Manley – DD-940. Destroyer

USS Neosho – AO-143. Oiler

USS New – DD-818. Destroyer

USS Preserver – ARS-8. Salvage ship – Another logical name.

USS Waldron – DD-699. Destroyer

Pacific:

USS George MacKenzie – DD-836. Destroyer

USS Goldsborough – DDG-20. Guided missile destroyer

USS Chipola – AO-63. Oiler

USS Leonard F. Mason – DD-852. Destroyer – Renown is now just a bit closer.

USS Taylor – DD-468. Destroyer.

GEMINI-TITAN 6 (Scrubbed)

Atlantic:
USS Wasp – CVS-18. Prime recovery carrier
USS Paiute – ATF-159. Fleet ocean tug
USS Fearless – MSO-442. Minesweeper ocean
USS Fidelity – MSO-443. Minesweeper ocean
USS Basilone – DD-824. Destroyer
USS Jonus Ingram – DD-938. Destroyer
USS Massey – DD-778. Destroyer
USS Aucilla – AO-56. Oiler
USS Stickell – DD-888. Destroyer
USS Noa – DD-841. Destroyer
USS Richard E. Kraus – DD-849. Destroyer

Pacific:
USS Cochrane – DDG-21. Guided missile destroyer
USS Renshaw – DD-499. Destroyer
USS Rupertus – DD-851. Destroyer
USS George K. MacKenzie – DD-836. Destroyer

GEMINI-TITAN 6A – WALLY SCHIRRA AND TOM STAFFORD DEC. 15 - 17, 1965

Atlantic:
USS Wasp – CVS-18. Prime recovery carrier
USS Ability – MSO-519. Minesweeper ocean
USS Aucilla – AO-56. Oiler
USS Meredith – DD-890. Destroyer
USS Paiute – ATF-159. Fleet ocean tug
USS Power – DD-839. Destroyer
USS Waccamaw – AO-109. Oiler
USS Waldron – DD-699. Destroyer
USS Joseph P. Kennedy – DD-850. Destroyer

Pacific:
USS Cochrane – DDG-21. Destroyer
USS George MacKenzie – DD-836. Destroyer
USS Renshaw – DD-449. Anti-sub destroyer

USS Rupertus – DD-851. Destroyer
USS Ponchatoula – AO-148. Oiler

GEMINI-TITAN 7 – FRANK BORMAN AND JIM LOVELL DEC. 4 - 18, 1965

Atlantic:
USS Wasp – CVS-18. Prime recovery ship
USS Ability – MSO-519. Minesweeper ocean
USS Aucilla – AO-56. Oiler
USS Joseph P. Kennedy – DD-850. Destroyer
USS Meredith – DD-890. Destroyer
USS Paiute – ATF-159 Fleet ocean tug
USS Power – DD-839. Destroyer
USS Waccamaw – AO-109. Oiler
USS Waldron – DD-699. Destroyer

Pacific:
USS Cochrane – DDG-21. Guided missile destroyer
USS George MacKenzie – DD-836. Destroyer
USS Renshaw – DDE-499. Anti-sub destroyer
USS Rupertus – DD-851. Destroyer
USS Ponchatoula – AO-148. Oiler

APOLLO-SATURN 201 (Unmanned) FEB. 26, 1966

Atlantic:
USS Boxer – LPH-4. Landing Platform Helicopter – Prime recovery carrier
USS Kankakee – AO-39. Oiler
USS Salinan – ATF-161. Fleet ocean tug
USS Fidelity – MSO-443. Minesweeper ocean
USS Beale – DD-471. Destroyer
USS Waller – DD-466. Destroyer
USS Bordelon – DD-881. Destroyer
USS Kaskaskia – AO-27. Oiler
USS Turner – DD-834. Destroyer
USS Robert L. Wilson – DD-847. Destroyer
USS Jonus Ingram – DD-938. Destroyer
USS Waldron – DD-699. Destroyer

GEMINI-TITAN 8 – NEIL ARMSTRONG AND ED SCOTT
MARCH 16-17, 1966

Atlantic:
USS Boxer – LPH-4. Prime recovery carrier
USS Caloosahatchee – AO-98. Oiler
USS Charles P. Cecil – DD-835. Radar picket destroyer
USS Fidelity – MSO-443. Minesweeper ocean
USS Goodrich – DD-831. Radar picket destroyer
USS Myles C. Fox – DD-829. Destroyer
USS Noa – DD-841. Destroyer
USS Paiute – ATF-159. Fleet ocean tug

Pacific:
USS Leonard F. Mason – DD-852. Destroyer – ACTUAL recovery ship.
USS Cochrane – DDG-21. Guided missile destroyer
USS George K. MacKenzie – DD-836. Destroyer
USS Hassayampa – AO-145. Oiler

GEMINI-TITAN 9A – TOM STAFFORD AND
EUGENE CERNAN
JUNE 3 - 6, 1966

Atlantic:
USS Wasp – CVS-18. Prime recovery carrier yet again
USS Bordelon – DD-881. Radar picket destroyer
USS Chikaskia – OA-54. Oiler
USS McCaffery – DDE-860. Anti-sub destroyer
USS Nimble – MSO-459. Minesweeper ocean
USS Opportune – ARS-41. Salvage ship
USS Papago – ATF-160. Fleet ocean tug
USS Robert L. Wilson – DD-847. Destroyer
USS Sabine – AO-25. Oiler
USS William C. Lawe – DD-763. Destroyer

Pacific:
USS Epperson – DD-719. Destroyer
USS George K. MacKenzie – DD-836. Destroyer
USS Hassayampa – AO-145. Oiler
USS Rupertus – DD-851. Destroyer

GEMINI-TITAN 10 – JOHN YOUNG AND
MICHAEL COLLINS
JULY 18 - 21, 1966

Atlantic:
USS Guadalcanal – LPH-7. Prime recovery carrier
USS Allen M. Sumner – DD-692. Destroyer
USS Norris – DD-859. Anti-sub destroyer
USS Opportune – ARS-41. Salvage ship
USS Severn – AO-61. Oiler
USS William C. Lawe – DD-763. Destroyer

Pacific:
USS Benjamin Stoddert – DDG-22. Guided missile destroyer
USS Collet – DD-730. Destroyer
USS DeHaven – DD-727. Destroyer
USS Kawishiwi – AO-146. Oiler

APOLLO-SATURN – 202 (Unmanned)
AUG. 25, 1966

Atlantic:
USS Opportune – ARS-41. Salvage ship
USS James C. Owens – DD-776. Destroyer
USS Robert A. Owens – DD-827. Destroyer
USS Salamonie – AO-26. Oiler
USS Chikaskia – AO-54. Oiler

Pacific:
USS Hornet – CVS-12. Prime recovery carrier
USS O'Bannon – DD–450. Destroyer
USS Sproston – DD-577. Destroyer
USS Benjamin Stoddert – DDG-22. Destroyer

GEMINI-TITAN 11 – PETE CONRAD AND DICK GORDON
SEPT. 12-15, 1966

Atlantic:
USS Guam – LPH-9. Prime recovery carrier
USS Forest Royal – DD-872. Destroyer
USS McCaffery – DD-860. Anti-sub destroyer

USS Nimpuc – ATF-157. Fleet ocean tug
USS Severn – AO-61. Oiler
USS Wallace L. Lind – DD-703. Destroyer
Pacific:
USS Kawishiwi – AO-146. Oiler
USS Mansfield – DD-728. Destroyer
USS Theodore Chandler – DD-717. Destroyer
USS O'Brien – DD-725. Destroyer

TITAN 3C/HST (Unmanned)
NOV. 3, 1966

Atlantic:
USS Lasalle – LPD-3. Amphibious transport dock – Prime recovery ship.
USS Fort Snelling – LSD-30. Landing ship dock
USS Aucilla – AO-56. Oiler

GEMINI-TITAN 12 – JIM LOVELL AND BUZZ ALDRIN
NOV. 11-15, 1966

Atlantic:
USS Wasp – CVS-18. Prime recovery carrier – Her fifth and last outing.
USS Canisteo – AO-99. Oiler
USS Charles H. Roan – DD-853. Destroyer
USS Joseph P. Kennedy – DD-850. Destroyer
USS Kankakee – AO-39. Oiler
USS Lloyd Thomas – DD-764. Destroyer
USS Preserver – ARS-8. Salvage ship
Pacific:
USS Hollister – DD-788. Destroyer
USS Joseph Strauss – DDG-16. Guided missile destroyer
USS Kawishiwi – AO-146. Oiler
USS Ozbourn – DD-846. Destroyer

That did it for Gemini. Enter Apollo, the system that would take man to the Moon.

APOLLO-SATURN 204 (APOLLO 1)
JAN. 27, 1967

This was, of course, the launch pad disaster which took the lives of astronauts Gus Grissom, Ed White and Roger Chaffee at the Cape. It stopped the space program in its tracks for many months while modifications were made. Despite the tragic fact that those men died in a launch pad test, there was a recovery fleet in both oceans.
Atlantic:
USS Essex – CVS-9. Prime recovery ship
USS Cony – DD-508. Destroyer
USS Dupont – DD-941. Destroyer
USS Kankakee – AO-39. Oiler
USS Lorain County – LST-1177. Landing ship tank
USS Salinan – ATF-161. Fleet ocean tug
Pacific:
USS Kawishiwi – AO-146. Oiler
USS Philip – DD-498. Destroyer
USS Radford – DD-446. Destroyer

APOLLO-4 (Unmanned)
NOV. 9, 1967

Atlantic:
USS Austin – LPD-4. Amphibious transport dock
USS Sabine – AO-25. Oiler
USS Joseph P. Kennedy – DD-850. Destroyer
USS York County – LST-1175. Landing ship tank
USS Hoist – ARS-40. Salvage ship
Pacific:
USS Bennington – CVS-20. Prime recovery ship
USS Carpenter – DD-825. Destroyer

APOLLO-5 (Unmanned)
JAN. 22, 1968

Atlantic:
USS Paiute – ATF-159. Fleet ocean tug

APOLLO–6 (Unmanned)
APRIL 4, 1968

Atlantic:
USS Austin – LPD-4. Amphibious transport dock
USS Chikaskia – AO-54. Oiler
USS York County – LST-1175. Landing ship tank
USS Dupont – DD-941. Destroyer
USS Opportune – ARS-41. Salvage ship

Pacific:
USS Okinawa – LPH-3. Prime recovery ship
USS Carpenter – DD-825. Destroyer

APOLLO-SATURN 7 – WALLY SCHIRRA, DON EISELE AND WALT CUNNINGHAM
OCT. 11 - 22, 1968

Atlantic:
USS Essex – CV-9. Prime recovery carrier – An entire class named after her.
USS Arneb – LKA-56. Amphibious cargo ship
USS Cambria – LPA-36. Amphibious transport
USS Paiute – ATF-159. Fleet ocean tug

Pacific:
USS Cochrane – DDG-21. Guided missile destroyer
USS Henry Tucker – DD-875. Destroyer
USS Nicholas – DD-449. Anti-sub destroyer
USS Ponchatoula – AO-148. Oiler
USS Rupertus – DD-851. Destroyer

APOLLO-SATURN 8 – FRANK BORMAN, JIM LOVELL AND BILL ANDERS
THE CHRISTMAS MISSION
DECEMBER 21 - 27, 1968

Atlantic:
USS Chuckawan – AO-100. Oiler
USS Francis Marion – LPA-249. Amphibious transport

USS Guadalcanal – LPH-7. Carrier
USS Rankin – LKA-103. Amphibious cargo ship
USS Salinan – ATF-161. Fleet ocean tug
USS Sandoval – LPA-194. Amphibious transport

Pacific:
USS Yorktown – CVS-10. Prime recovery carrier
USS Arlington – AGMR-2. Communications relay ship
USS Chipola – AO-63. Oiler
USS Cochrane – DDG-21. Guided missile destroyer
USS Nicholas – DD-449. Anti-sub destroyer
USS Rupertus – DD-851. Destroyer

APOLLO-SATURN 9 – JIM McDIVITT, DAVE SCOTT AND RUSTY SCHWEIKART
MARCH 3 - 13, 1969

Atlantic:
USS Guadalcanal – LPH-7. Prime recovery carrier
USS Algol – LKA-54. Amphibious cargo ship
USS Paiute – ATF-159. Fleet ocean tug

Pacific:
USS Cochrane – DDG-21. Guided missile destroyer
USS Leonard F. Mason – DD-852. Destroyer
USS Nicholas – DD-449. Anti-sub destroyer

Notice how these recovery fleets are suddenly shrinking in size.

APOLLO-SATURN 10 – TOM STAFFORD, JOHN YOUNG AND EUGENE CERNAN
MAY 18 - 26, 1969

Atlantic:
USS Chilton – LPA-38. Amphibious transport
USS Ozark – MCS-2. Mine countermeasures support ship
USS Rich – DD-820. Destroyer
USS Salinan – ATF-161. Fleet ocean tug

Pacific:
USS Princeton – LPH-5. Prime recovery carrier – Entered Naval service as an Essex class aircraft carrier but finished her career as a helicopter carrier

USS Arlington – AGMR-2. Communications relay ship
USS Carpenter – DD-825. Destroyer

APOLLO-SATURN 11 – NEIL ARMSTRONG, BUZZ ALDRIN AND MIKE COLLINS
JULY 16 - 24, 1969
THE PRIZE!

Atlantic:
USS New – DD-818. Destroyer
USS Ozark – MCS-2. Mine countermeasures support ship
USS Salinan – ATF-161. Fleet ocean tug

Pacific:
USS Arlington – AGMR-2. Communications relay ship
USS Hornet – CVS-12. Prime recovery carrier for a premium mission
USS Hassayampa – AO-145. Oiler
USS Goldsborough – DDG-20. Guided missile destroyer

APOLLO-SATURN 12 – PETE CONRAD, DICK GORDON AND ALAN BEAN
NOV. 14 - 24, 1969

Atlantic:
USS Austin – LPD-4. Amphibious transport dock – Very ship-like, really.
USS Escape – ARS-6. Salvage ship
USS Hawkins – DD-873. Destroyer

Pacific:
USS Hornet – CVS-12. Prime recovery carrier
USS Joseph J. Strauss – DDG-16. Guided missile destroyer

APOLLO-SATURN 13 – JIM LOVELL, FRED HAISE, AND JOHN SWIGERT
APRIL 11- 17, 1970
HOUSTON – WE HAVE A PROBLEM!

Atlantic:
USS Bordelon – DD-881. Radar picket destroyer
USS Forest Royal – DD-872. Destroyer

USS New – DD-818. Destroyer
USS William C. Lawe – DD-763. Destroyer
USS Escape – ARS-6. Salvage ship

Pacific:
USS Iwo Jima – LPH-2. Prime recovery carrier – Most of the LPH's (not all) were named after major Pacific island WWII battles.
USS Benjamin Stoddert – DDG-22. Guided missile destroyer
USS Granville S. Hall – YAG-40. Miscellaneous auxiliary service craft
USS Kawishiwi – AO-146. Oiler

APOLLO-SATURN 14 – ALAN SHEPARD, ED MITCHELL AND STUART ROOSA
JAN. 31 – FEB. 9, 1971

Atlantic:
USS Hawkins – DD-873. Destroyer
USS Paiute – ATF-159. Fleet ocean tug
USS Spiegel Grove – LSD-32. Landing ship dock

Pacific:
USS New Orleans – LPH-11. Prime recovery carrier
USS Carpenter – DD-825. Destroyer
USS Ponchatoula – AO-148. Oiler

APOLLO-SATURN 15 – DAVE SCOTT, JIM IRWIN AND AL WORDEN
JULY 26 – AUG. 7, 1971

Atlantic:
USS Austin – LPD-4. Amphibious transport dock
USS Salinan – ATF-161. Fleet ocean tug

Pacific:
USS Okinawa – LPH-3. Prime recovery carrier
USS Kawishiwi – AO-146. Oiler

APOLLO-SATURN 16 – JOHN YOUNG, TOM MATTINGLY
AND CHARLES DUKE
APRIL 16 - 27, 1972

Atlantic:
USS Alacrity – MSO-520. Minesweeper ocean
USS Opportune – ARS-41. Salvage ship
USS Exploit – MSO-440. Minesweeper ocean
Pacific:
USS Ticonderoga – CVS-14. Prime recovery carrier
USS Goldsborough – DDG-20. Guided missile destroyer
USS Ponchatoula – AO-148. Oiler

APOLLO-SATURN 17 – GENE CERNAN, RON EVANS
AND HARRISON SCHMITT
DEC. 7 - 19, 1972
THE LAST HURRAH

USS Ticonderoga – CVS-14. Prime recovery carrier
USS Camden – AOE-2. Replenishment ship – In other words, an
   oiler.
USS Recovery – ARS-43. Salvage ship
USS Saginaw – LST-1188. Landing ship tank

Goodbye to Apollo and six great lunar landings. On to Skylab and
an historic ending to the splashdown era.

SKYLAB 2 – PETE CONRAD, PAUL WEITZ
AND JOE KIRWIN
MAY 25 - JUNE 22, 1973
Better living conditions pave the way for longer missions.
USS Ticonderoga – CVS-14. Prime recovery carrier
USS Escape – ARS-6. Salvage ship
USS Grapple – ARS-53. Salvage ship

SKYLAB 3 – ALAN BEAN, JACK LOUSMA
AND OWEN GARRIOTT
JULY 28 – SEPT. 25, 1973
USS New Orleans – LPH-11. Prime recovery ship
USS Escape – ARS-6. Salvage ship
USS Grapple – ARS-53. Salvage ship

SKYLAB 4 – GERALD CARR, ED GIBSON
AND BILL POGUE
NOV. 16, 1973 – FEB. 8, 1974
THE LONGEVITY CHAMPION

USS New Orleans – LPH-11. Prime recovery carrier
USS Opportune – ARS-41. Salvage ship

APOLLO-SOYUZ TEST PROJECT (ASTP) – TOM STAFFORD,
DEKE SLAYTON, VANCE BRAND
AND SOVIET COSMONAUTS ALEXEI LEONOV
AND VALERY KUBASOV
JULY 15 - 24, 1975
END OF SPLASHDOWNS

USS New Orleans – LPH-11. Prime recovery ship – Good thing.
Looks like there was nobody else around.

*Note – On many manned recoveries, ships which remained in or close
   to their home ports are counted among those at least able to participate
   if needed. This is why you see occasional differences between this list
   and written accounts appearing later in the book.

The book you have begun reading remains a testament to all those ships and units involved in MANNED RECOVERIES but when we were presented with a list showing the ships involved in both manned and unmanned missions it was easy to decide to include them all. Why not? They did their jobs and had no say as to whether the spacecraft they were protecting and ultimately looking for were empty or occupied by human beings....or even a few chimpanzees. We ended up counting 236 different ships but many were involved in more than one recovery, some as many as the SS Paiute (9 missions), so the final tally was more like 459 counting the repeaters. That's an incredible record but... it's an incredible Navy.

And just in case the names of certain ships seem to have a certain "ring" to them, such as Hassayampa, Chukawan, Chikaskia, Kankakee and so forth, be advised that most (but not all) of the fleet oilers on this list were named after American Indian tribes.

Author's note: This list was constructed from NASA records that came to us from two separate NASA offices and there were differences between the two. What you have here is the result of a lot of checking and double-checking that we believe led to a list that is as close to 100% correct as we could possibly make it and almost certainly represents the first time it has been published in hard cover.

Ship name	# of Missions	Ship name	# of Missions	Ship name	# of Missions
Abbot DD-629	1	Blandy DD-943	3	Cone DD-866	3
Ability MSO-517	3	Bordelon DD-881	5	Conway DD-507	2
Adroit MSO-509	1	Borie DD-704	3	Cony DD-508	5
Affray MSO-511	1	Boston CA-69	1	Davis DD-937	1
Agile MSO-421	1	Boxer LPH-4	2	Decatur DDG-31	3
Alacrity MSO-520	3	Bristol DD-857	1	DeHaven DD-727	2
Algol LKA-54	1	Brownson DD-868	1	Dewey DLG-14	1
Allen M. Sumner DD-692	1	Bulwark MSO-425	1	Diligence WMEC-616	1
Antietam CVS-36	1	Caloosahatchee AO-98	1	Donner LSD-20	4
Arlington AGMR-2	3	Cambria LPA-36	1	Douglas H. Fox DD-779	1
Arneb LKA-56	1	Camden AOE-2	1	Duncan DDR-874	1
Aucilla AO-56	4	Canisteo AO-99	1	DuPont DD-941	3
Ault DD-698	1	Carpenter DD-825	4	Dyess DDR-880	1
Austin LPD-4	4	Casa Grande LSD-13	1	Eaton DD-510	1
Avenge MSO-423	1	Charles Adams DDG-13	1	Elokomin AO-55	1
Bache DD470	1	Charles H. Roan DD-853	1	English DD-696	1
Barry DD-933	3	Charles P. Cecil DD-835	1	Epperson DDE-719	3
Barton DD-722	1	Charles S. Sperry DD-697	3	Escape ARS-6	7
Basilone DD-824	2	Chikaskia AO-54	4	Essex CVS-9	2
Beale DD-471	2	Chilton LPA-38	1	Eugene A. Greene DD-71	3
Bearss DD-654	1	Chipola AO-63	3	Exploit MSO-440	3
Beatty DD-756	2	Chukawan AO-100	4	Exultant MSO-441	1
Benjamin Stoddert DDE-22	3	Cochrane DDG-21	7	Farragut DLG-6	1
Bennington CVS-20	1	Collet DD-730	1	Fearless MSO-442	1
Bigelow DD-942	3	Compton DD-705	2	Fidelity MSO-443	4

Ship name	# of Missions
Fiske DD-842	1
Fletcher DD-445	2
Forrest Royal DD-872	3
Forrestal CVA-59	2
Fort Mandan LSD-21	1
Fort Snelling LSD-30	1
Francis Marion LPA-249	1
Frank Knox DD-742	1
Fred T. Berry DDE-858	2
Furse DDR-882	3
Gainard DD-706	1
George K. MacKenzie DD-836	6
Glennon DD-840	2
Goldsborough DDG-20	4
Goodrich DDR-831	2
Granville S. Hall YAG-40	1
Grapple ARS-53	2
Guadalcanal LPH-7	3
Guam LPI I-9	1
Gyatt DD-712	1
Hailey DD-556	1
Hank DD-702	1
Harold J. Ellison DD-864	2
Harwood DD-861	2
Hassayampa AO-145	3
Hawkins DD-873	4
Haynesworth DD-700	1
Henley DD-391	1
Henry Tucker DD-875	1
Higbee DD-806	1
Hoist ARS-40	5
Holder DD-819	1
Hollister DD-788	1
Hornet CVS-12	3
Hugh Purvis DD-709	1
Hunt DD-674	1
Hyman DD-732	2
Independence CV- 62	1

Ship name	# of Missions
Ingraham DD-694	1
Intrepid CVS-11	2
Iwo Jima LPH-2	1
James C. Owens DD-776	2
John A. Bole DD-755	1
John Paul Jones DD-32	2
John R. Pierce DD-753	1
John W. Thomason DD-760	1
John W. Weeks DD-701	1
John Willis DE-1027	1
Jonus Ingram DD-938	2
Joseph J. Strauss DDG-16	2
Joseph P. Kennedy DDD-850	6
Kankakee AO-39	4
Kaskaskia AO-27	2
Kawishiwi AO-146	7
Kearsarge CVS-33	1
Kenneth D. Bailey DD-713	1
Laffey DD-724	1
Lake Champlain CVS-39	5
Lasalle LPD-3	1
Leonard F. Mason DD-852	4
Lloyd Thomas DD-764	1
Lofberg DD-759	1
Lorain County LST-1177	1
Lowry DD-770	2
Manley DD-940	3
Mansfield DD-728	2
Massey DD-778	2
McCaffery DDE-860	2
Meredith DD-890	2
Moale DD-693	1
Mullinix DD-944	1
Murray DD-576	1
Myles C. Fox DD-829	2
Neosho AO-143	1
New DD-818	3
New Orleans LPH-11	4

Ship name	# of Missions
Newman K. Perry DD-883	2
Nicholas DD-449	3
Nimble MSO-459	2
Nipmuc AFT-157	2
Noa DD-841	3
Norfolk DL-1	1
Norris DDE-859	2
Notable AM-267	1
O'Bannon DDE-450	3
O'Brien DD-725	1
Observer AM-461	1
O'Hare DD-889	1
Okinawa LPH-3	2
Opportune ARS-41	8
Orleck DD-886	1
Ozark MCS-2	2
Ozbourn DD-846	1
Paiute ATF-159	9
Papago ATF-160	1
Perry DD-844	2
Philip DD-498	2
Plymouth Rock LSD-29	1
Ponchatoula AO-148	6
Power DD-839	3
Preserver ARS-8	2
Princeton LPH—5	1
Purdy DD-734	1
Putnam DD-757	1
Radford DDE-446	2
Randolph CVS-15	2
Rankin LKA-103	1
Recovery ARS-43	5
Remey DD-688	1
Renshaw DD-499	3
Rich DD-820	3
Richard E. Kraus DD-849	1
Robert A. Owens DD-827	3
Robert H. McCard DD-822	2

Ship name	# of Missions	Ship name	# of Missions	Ship name	# of Missions
Robert L. Wilson DD-847	3	Steinaker DD-863	1	Waccamaw AO-109	2
Robinson DD-562	1	Stickell DD-888	1	Wadleigh DD-689	1
Rooks DD-804	1	Stormes DD-780	3	Waldron DD-699	4
Rupertus DD-851	7	Stribling DD-876	1	Walker DDE-517	1
Sabine AO-25	2	Sturdy MSO-494	2	Wallace L. Lind DD-703	2
Saginaw LST-1188	1	Sullivans DD-537	1	Waller DD-466	2
Salamonie AO-26	1	Swerve MSO-495	2	Warrington DD-843	1
Salinan ATF-161	6	Taussig DD-746	1	Wasp CVS-18	7
Sandoval LPA-194	1	Taylor DD-468	1	Willard Keith DD-775	1
Sarsfield DD-837	3	Theodore Chandler DD-717	1	William C. Lawe DD-763	3
Severn AO-61	2	Ticonderoga CVS-14	3	Witek EDD-848	2
Skill MSO-471	1	Turner DD-834	2	Wren DD-568	1
Soley DD-707	1	Valley Forge LPH-8	1	York County LST-1175	2
Spiegel Grove LSD-32	2	Vesole DD-878	1	Yorktown CVS-10	1
Sproston DD-577	1	Vigilant WPC-617	1	Zellars DD-777	1
Stalwart AM-493	1	Vogelgesang DD-862	2		

The Champions

USS New Orleans – 4 Recoveries

USS Wasp – 5 Recoveries

USS Ticonderoga – 3 Recoveries

THE HELICOPTERS

If there was one company, one aircraft manufacturer, that dominated the manned recoveries it was Sikorsky, the famed helicopter maker whose huge plant sits alongside Connecticut's Housatonic River in Stratford and looks out on the equally famed Merritt Parkway.

No doubt a few other manufacturer's helicopters got into the recovery act here and there, but from the moment Alan Shepard splashed down in his tiny Mercury Redstone-3 capsule (Freedom 7) in 1961 until the Apollo portion of Apollo-Soyuz in 1975...Sikorsky machines ruled the roost whether in the Atlantic or the Pacific.

From the outset it was either one Sikorsky model or another starting with the S-58 on the Alan Shepard mission. The Navy was still enjoying their use well into the decade of the 1960's.

Yes, there were other makes and models such as the Piasecki H-25 involved with the recovery of astronaut John Glenn when his Mercury Atlas 6 (Friendship 7) capsule came down following Glenn's first-American-in-orbit flight early in 1962, and the destroyer Noa was the closest to the astronaut at splashdown. A Piasecki took Glenn from the Noa back to the primary recovery carrier USS Randolph.

But you have to look high and low to find many other helicopters not bearing the proud name Sikorsky on them. Starting with Mercury-Atlas 7 (Aurora 7), the Scott Carpenter mission, the SH-3 stepped to the forefront; and, with a few exceptions, became the predominant helicopter right up until the final splashdown.

As carefully explained to me by Sikorsky Humanitarian Communication's Manager, Bill Tuttle, one summer day in 2003, the great H-3 series manifested itself in many forms; but under the Navy's

Right: From one living legend to another. It had to be a mutual admiration society. Igor Sikorsky shakes hands with Neil Armstrong. Neil's comment below the photo is not entirely readable. He wrote, "To Igor Sikorsky – with admiration and respect of a junior exponent of vertical take-off and landing."

wing, it was always preceded by that SH and was proudly called "Sea King." Not only did H-3's fly for many nations for decades, but they were actually built "off shore" in considerable numbers – in Japan, for instance, by Mitsubishi. From Tuttle I learned that over 1,200 H-3s were built – 791 at the Sikorsky plant in Connecticut, the rest overseas. Manufacture at Sikorsky began in 1959 and ended in 1980.

While Navy helicopters such as the HS-3 number 66 gained international acclaim as the primary recovery chopper for Apollo 11 and four other Apollo missions, no model or "variant" is more familiar to millions of Americans than VH-3D. It uses the radio call sign Marine One and is painted an easy-on-the-eyes olive drab with white top. It takes our Presidents to Camp David, Air Force One, and elsewhere. If it's presidential…. it's a VH.

Most of the services and nations which bought and benefited from their use of the H-3 helicopters have moved on to succeeding models which still carry the name Sikorsky although many H-3s remain in use – primarily with nations such as Canada, Brazil and Malaysia.

Once again, the man whose name is practically a synonym for helicopters, Igor Sikorsky (right foreground), meets and greets a trio of modern-day air aces – the Apollo 16 trio of (from left) John Young, Charles Duke and Tom Mattingly.

We have written that there was more than one helicopter manufacturer's aircraft involved in the space program besides the overwhelming presence of Sikorsky. We found one... just one. Piasecki. An early Piasecki, the PV-2, was just the second successful helicopter to fly in the U.S. on April 11, 1943.

The one shown here, however, debuted in the late 1950s and was called the HUP-2, designed specifically for search and rescue. When John Glenn overshot his prime recovery carrier, the USS Randolph, it was a Piasecki HUP, also known as the H-25, which hoisted the capsule to the deck of the destroyer USS Noa, and about three hours later, took him over to the Randolph. (Courtesy of www.aviation.army.mil)

The original Flying Tigers wrote their own history in the early dark days of World War II and they flew the legendary P-40. The Flying Tigers who took part in the first phase of this nation's space program were not even organized until 1951 and they were flying helicopters, not fighter planes.

HMM-262 was activated on September 1 of that year at Cherry Point, North Carolina and was originally known as a Marine Helicopter Transport Squadron. Despite the Korean War the unit remained on the east coast, became part of Marine Aircraft Group 26 (MAG) and was transferred to New River, North Carolina, in July of 1954, remaining there until 1966. That deployment would more than cover the squadron's involvement in Mercury.

Actually, the squadron was selected for its role in the space program in the late 1950's by NASA and put under the direction of the Project Mercury Recovery Force. In the years prior to Alan Shepard's sub-orbital flight in 1961, the squadron ranged everywhere up and down the east coast from NASA's Langley Field in Virginia to Maryland's Chincoteague Island and down to Cape Canaveral and back. They practiced recoveries with the astronauts and went out and recovered unmanned capsules even before NASA put a few chimpanzees in them. In short, they were extensively trained long before the nation was introduced to the original Mercury Seven.

When the big day came, the Tigers were ready. Here is a photo of their helicopter number 44 bringing astronaut Alan Shepard back to the prime recovery ship, the USS Lake Champlain.

HMR(L)-262's helicopter #44 (Sikorsky's S-58, Marine designation UH-34D), shown bringing the Mercury-Redstone 3 spacecraft to the deck of the carrier USS Lake Champlain. (Courtesy of NASA)

The device which hangs below the spacecraft was for impact reduction and preceded the flotation collar which was introduced by the time John Glenn became the first American to orbit the Earth. This retrieval was a success but, on the second sub-orbital, Gus Grissom's spacecraft had become flooded when the hatch blew open prematurely, giving the astronaut all he could handle to keep himself afloat. By the time the Marine UH-34D Seahorse came along to retrieve the spacecraft, it was already a full thousand pounds beyond the helicopter's lifting capacity, the capsule slipped the helo's grasp and went to the bottom of the Atlantic.

The squadron's most active involvement in the space program turned out to be their on-scene participation in the initial two Mercury sub-orbital missions. For the remaining four Mercury flights, the squadron, while not involved in the actual recoveries, was nonetheless on station at the end of each orbit, fully operational and equipped to do the job if called upon.

Lest there be confusion in anyone's mind as to the designation of HMM-262 or HMR (L)-262 it merely reflected a change in the description of their UH-34D Sikorsky helicopters as either Medium or Light. Still the same squadron.

Succeeding years found the squadron training and fine-tuning the helicopter doctrine utilized by the United States Marine Corps and which was put into practice in the spring of 1965 when it participated in the Dominican Republic intervention.

The next conflict, however, was the Vietnam War and there was no way HMM-262 was going to remain stateside. The squadron was reassigned to Marine Aircraft Group 36, 1st Marine Air Wing and took part in numerous operations across that country from aboard the USS Tripoli (LPH-10).

Following Vietnam, it was off to Hawaii and helicopter support for the 1st

Marine Amphibious Brigade followed by a deployment to the western Pacific in December, 1976, as a composite squadron where it was once more welcomed to the deck of the Tripoli. Before heading west however, the squadron garnered one of its first awards – the Commandant's Aviation Efficiency Award for accomplishments during the preceding year.

This is probably the entire HMR(L)-262 group which kicked off the first two recoveries. Unfortunately this picture was not accompanied by names. The six men in front are all pilots or co-pilots while the fifteen standing are all crewmembers.

In January of 1979, the Flying Tigers were awarded the highest grade ever achieved under the Marine Corps Combat Readiness Evaluation System for duties performed aboard the USS Belleau Wood (LHA-3) a spanking new amphibious assault ship. Soon afterwards, another first, as HMM-262 became the first helicopter squadron to deploy aboard the Navy's newest class of amphibious assault ships, the USS Tarawa (LHA-1) and then yet another first when the squadron operated jointly off the same deck with a detachment of AV-8A harriers, the so-called "jump jets," which would rise off a deck like a helicopter before taking off much like a conventional fixed-wing aircraft.

Into the 80's and the squadron would assist during the U.S. hostage crisis in Iran, embark on its fourth WESTPAC deployment aboard the USS Okinawa (LPH-3) and support Afghanistan contingency operations in the northern Arabian Gulf.

They were back aboard the Belleau Wood for WESTPAC deployment number five, a visit to Somalia after the departure of Russian forces there and a mission to Australia.

In August of 1983 another very high combat readiness evaluation award followed quickly by participation in a dual-pronged surface and air assault exercise at Iwo Jima.

January 1985. A squadron with a heart. Operation Handclasp brought 3,600 pounds of books to needy children in the Philippines and 13,000 pounds of vital material to a leper colony there.

The squadron's helicopters were upgraded in the summer of 1987, and the Flying Tigers found themselves back aboard Belleau Wood in March 1990. September 1992, still on the Belleau Wood for hurricane relief to the Hawaiian island of Kauai.

By 1993 the squadron was active in exercises in the Mariana Islands, Singapore and Australia, and off the coast of Okinawa. A year later HMM-262 helped celebrate the 50th anniversary of the Liberation of Guam and then, just for something different, squadron helicopters dropped a total of 255,000 pounds of water to help battle brush fires on Okinawa.

In 1996 the squadron was deeply involved in operations in the Philippines, Indonesia, Malaysia, Singapore, Thailand and South Korea. HMM-262's 45th birthday was noted with a formation, cake, liberty, and a picnic and softball game.

Once again, the USS Belleau Wood became the squadron's take-off point for exercises Cobra Gold 97, Foal Eagle 97, Valiant Usher 98 and Cobra Gold 98 as well as operations in all the western Pacific nations noted previously.

In January of 2000, the Flying Tigers of HMM-262 became the Combat Air Element of the 31st Marine Expeditionary Unit and by February 2002, they replaced HMM-262 as the 31st MEU's Air Combat Element.

—HC-1 (AND EVENTUALLY HC-2)— HELICOPTER COMBAT SUPPORT SQUADRON

"Fleet Angels"

A subtitle might be "Oh what a tangled web we weave" and while there is no intent to deceive in this history, it does get a bit tangled.

Life for the United States Navy's first operational helicopter squadrons, HU-1 and HU-2, began on April 1, 1948, at the Naval Air Station in Lakehurst, New Jersey. They were designed to provide utility services to ships of both the Atlantic and Pacific Fleets which goes along to explain how an early detachment became involved in the John Glenn mission in the Atlantic, while the other became the unchallenged champion of Pacific recoveries.

They flew from both aircraft carriers and smaller air-capable ships – an important point because as helicopter versatility gained recognition, understandably, new squadrons, each with more specialized functions were split off from

HELSUPPRON ONE

these two. HU-2 would remain an east coast unit while HU-1 was soon on its way to the Naval Auxiliary Air Station at Imperial Beach, California, where it became HC-1 in the mid 1960s and later proceeded to reel off an impressive string of manned recoveries.

Prior to any involvement in the space program, HU-1 would establish an outstanding reputation for air-sea rescue during the Korean conflict in the early 1950s. Its pilots and air crewmen pioneered new techniques of personnel rescue from behind enemy lines, picking up a Presidential Unit Citation and the Congressional Medal of Honor for one of its pilots, LTJG John Koelsch.

In 1955 HU-2 helicopter crews saved 500 civilians from severe flooding following a hurricane, which swept through Pennsylvania. Seven years later it was HU-2 again, this time rescuing some 1800 residents along the southern New Jersey coast after a crippling late winter storm.

Just a month earlier, a unit of HU-2 known as Detachment 36 and operating from the back-up recovery ship, the destroyer NOA (DD-841), raced in to retrieve America's first orbiting astronaut, John Glenn, on February 20, 1962, when Glenn overshot the primary recovery carrier Randolph (CVS-15) by more than a hundred miles.

HU-2 would become part of the air arm of the massive fleet in the Caribbean during the Cuban Missile crisis later that year. By the summer of 1965 HU-2 had become HC-2.

By the time the 1970's rolled around, HU-1 had been redesignated HC-1 for Helicopter Combat Support Squadron, which more accurately described its regular mission. However, that did not stop the Navy from putting the squadron to work in a string

of manned recoveries in the Pacific…. Apollo 15, 16 and 17 and Skylab II, III and IV.

HC-2 would continue its Atlantic operations right on into September of 1977 when it was disestablished and re-emerged a decade later, rebuilt with detachments of three different squadrons, and continuing to earn further prestigious awards for operational safety and battle efficiency.

HC-1 remains a west coast squadron with the Naval Air Station at North Island its most recent home, and they are still the "Fleet Angels."

HS-3

HELICOPTER ANTI-SUBMARINE 3

"Tridents"

Born (established) on June 18, 1952, at the Naval Air Facility, Elizabeth City, North Carolina. At first the squadron flew the Piasecki UH-25B but later transitioned to, and has remained with, Sikorsky aircraft starting with the H-19 and SH-34 helicopters then on to the familiar SH-3 Sea King for several decades and currently operating Sikorsky's SH-60F and HH-60H Seahawks.

Throughout its more than 50 year existence, HS-3's mission has been exactly what its name implies – anti-submarine detection and tracking and, in the event of war, the destruction of enemy subs. The squadron is one of 5 east coast operational squadrons with similar missions which include search and rescue.

HS-3 was introduced to the U.S. space program in August of 1962 when the unit's Commander J. M. Wondergrem picked up Lt. Commander M. Scott Carpenter from his Mercury spacecraft, Aurora 7, and brought him to the primary recovery carrier, USS Intrepid. Before the squadron's part in the space program was over, astronauts Gus Grissom and John Young (Gemini 3), Michael Collins (GT-10), Dick Gordon and Pete Conrad (GT-11) and Jim McDivitt, David Scott and Rusty Schweikart (Apollo 9), all ended their space journeys in Trident helicopters. A most impressive performance.

HS-4

"Black Knights"

Well, I guess if you're going to fly the helicopters involved in five Apollo missions including Apollo 11, and two before that and two more afterwards, then you logically would enjoy a very high status among your fellow helicopter squadrons.

As a matter of fact, it was for duties associated with their regular Naval assignments that made HS-4 the most decorated helicopter anti-submarine squadron in history. The Apollo recoveries only added to that illustrious record.

The squadron was commissioned in June of 1952 at the U.S. Naval Auxiliary Landing Field, Imperial Beach, California. The nickname "Black Knights" came about in 1961 when the squadron achieved around-the-clock and all-weather Anti-Submarine Warfare (ASW) capability. Yes, the nights were often quite black.

HS-4 further added to its growing reputation with the rescue of 24 downed airmen under hostile conditions during the 1966 Gulf of Tonkin operations off Vietnam and joined the USS Yorktown as part of a task force deployed to the Sea of Japan during the USS Pueblo crisis in 1968.

With that sort of a track record in place as NASA set its sights on the Moon, I would have insisted, had I been an astronaut at the time, on having the men and helicopters of HS-4 pick me up in the Pacific. Wouldn't you? Demand the best.

We will attend to the Black Knights impressive list of awards momentarily but first let us turn to the raison d'etre for this book – manned spacecraft and astronaut recovery. Be advised that HS-4 was there for the recoveries of Apollo 8 (the so-called Christmas mission of 1968), Apollo 10 (orbiting the

Moon), Apollo 11 and 12 (both lunar landings) and Apollo 13 (close…but no cigar and lots of white knuckles).

In that process Helicopter 66 got a lot of television and print media time and, let's face it, became world famous. Nice number – 66 – and so easy to remember.

As the decade of the 60's ended, HS-4 left the space program behind and returned to more traditional assignments including a couple of firsts. The deployment aboard the USS Kitty Hawk (CV-63) in 1973 was the first west coast deployment for the squadron and lasted for a decade following which the Black Knights embarked on the nuclear carrier USS Carl Vinson (CVN-70) for that ship's inaugural around-the-world cruise.

Throughout HS-4's recovery experience and on into the early 1990's the squadron flew the tried and tested Sikorsky Sea King SH-3H. In 1991, however, the unit transitioned to the H-60 Sikorsky Sea Hawk helicopters still in wide use today and returned to the USS Kitty Hawk (now CVW-15) as part of Carrier Air Wing Fifteen. In January 1995, HS-4 became part of the Carrier Air Wing Fourteen team.

In 2002 and 2003 the Black Knights were part of a world record western Pacific deployment aboard the USS Abraham Lincoln (CVN-72), during which their accomplishments included military engagement with Pakistan, detachments in Kuwait and CSAR training in Australia, all in support of Operation Southern Watch and Operation Enduring Freedom. During Operation Iraqi Freedom, HS-4 supported Navy SEAL, British Commando and Polish GROM forces in Kuwait and Iraq, and were the primary force protection for the USS Abraham Lincoln Battle Group.

Now…as to those awards. HS-4 is a seven-time winner of the Battle 'E' Efficiency Award, six-time winner of the Chief of Naval Operations Safety Award, five-time winner of the Capt. Arnold J. Isbell Award for ASW Excellence and five-time winner of the Sikorsky Excellence in Maintenance Award. They also received the Admiral Arleigh Burke Fleet Trophy and the Silver Anchor for excellence in personnel programs. Take a bow Knights…a big one.

HS-5

HELICOPTER ANTI-SUBMARINE SQUADRON 5

"Nightdippers"

This anti-submarine unit was brought to life on January 3, 1956, at the Naval Air Station in Key West, Florida, starting out with the Sikorsky HSS-1 Sea Horse helicopter.

As the 60's began the unit transitioned to the next generation Sikorsky, the ubiquitous SH-3A Sea King and used that helicopter in Anti-Submarine Warfare exercises as well as several spacecraft recoveries.*

In 1968 HS-5 exceeded 25,000 consecutive mishap free hours flying their SH-3A's and was presented with the Sikorsky Safety Award in recognition. It says a lot for both – the squadron and the helicopter maker.

In succeeding years HS-5 would garner a string of safety and maintenance awards and Meritorious Unit Commendations while serving in the Mediterranean from the decks of the nuclear-powered USS Dwight D. Eisenhower, then the Navy's newest nuclear powered carrier.

---

* This is a sad note. We tried very hard to do justice to the records of all units involved in the space program but some were so deeply imbedded that specifics became all but impossible. Such was the case with HS-5. References were frequent to "participation" in two Mercury spacecraft recoveries and "other" astronaut recovery missions in the mid 1960's which would have to mean Gemini. But every effort to pin down specific missions led to zero detail. No numbers...no names. We did uncover one direct tie-in and that was during the Apollo 7 splashdown in the Atlantic. We do not doubt the historic, but very generalized, references to earlier manned mission recoveries. We just wish the unit's historians had been more forthcoming with specifics.

The unit took part in Operation Desert Shield in the Red Sea in the summer of 1990 while those SH-3H helicopters kept chalking up high marks for endurance and reliability. In July of 1992 HS-5 was lifting off the decks of a new ship, the USS George Washington and more missions in the Mediterranean, the Adriatic Sea and the Arabian Gulf. As 1995 began the squadron welcomed a new Sikorsky helicopter, the H-60 Seahawk.

In late February 1998 the Nightdippers had still another new deck to fly off of – the USS John C. Stennis (CVN-74), spent four more months in the Arabian Gulf and then headed back to Jacksonville, Florida, in August of that year.

HS-6

HELICOPTER ANTI-SUBMARINE SQUADRON 6

"Indians"

Just to be different perhaps, this helicopter squadron was commissioned on board a carrier, the USS Princeton (CVS-37) on June 1, 1956, when the ship was midway between the islands of Formosa and Okinawa. One other factor, however, retained a familiarity to seemingly all over helicopter squadrons operating at the time. HS-6 began with Sikorsky's HO4S helicopter, transitioned to the same manufacturer's SH-34 two years later, and the SH-3 Sea King in 1962 and, pausing for upgrades now and then, worked with that aircraft right on into the 1990's. You could probably guess by now that the next helicopter to enter the picture was the Sikorsky H-60 Seahawk.

HS-6 helicopters operating off the deck of the USS Kearsarge (CVS-33) recovered astronauts Wally Schirra (Mercury 8) in 1962 and Gordon Cooper (Mercury 9) in 1963. Between those missions and the rescue of 16 downed

aviators in 1966 in support of American air strikes into Vietnam, the unit came to be respectfully referred to as the "World Famous Indians."

By 1974 HS-6 began its first regular assignment to an aircraft carrier, the USS Constellation (VC-64) and joined the Carrier Air Wing Eleven aboard the USS Enterprise (CVN-65) in 1981. The new H-60 helicopters arrived in 1990 and subsequently the squadron deployed on the USS Abraham Lincoln (CVN-72), the USS Kitty Hawk (CV-63) and the USS Carl Vinson (CVN-70) to the Western Pacific, the Indian Ocean and the Arabian Gulf. It remains a member of Carrier Air Wing 11 aboard the carrier Vinson.

HS-11
HELICOPTER ANTI-SUBMARINE SQUADRON 11

"Dragonslayers"

HS-11 was commissioned June 27, 1957, at the Naval Air Station, Quonset Point, Rhode Island, where it remained until mid-October 1973 when the Dragonslayers moved to their present home – the Naval Air Station in Jacksonville, Florida.

A very familiar picture emerges. HS-11 flew the Sikorsky SH-3 Sea King until 1994 before transitioning to the SH-60 Seahawk. It is currently part of Carrier Air Wing One and is embarked aboard the USS George Washington (CVN-73).

HS-11 took part in a series of astronaut recoveries in the 1960's — Gemini 4, 6 and 7 in 1965 and Gemini 12, late in 1966.

In its 45 year history the squadron has accumulated many awards for professionalism and excellence. Four Captain Arnold Jay Isbell trophies for ASW excellence, the coveted Admiral "Jimmy" Thach trophy for being the Navy's best carrier-based ASW squadron, and the Atlantic Fleet Golden Anchor Award in 1983, 1984 and 1995. Also numerous Battle Efficiency and Maintenance and Safety awards.

In March of 2001 HS-11 reached 30,000 flight hours and over 9 years of Class-A mishap-free flight. In September of 2001, our world changed and the war on terror began. HS-11 deployed on board the USS Theodore Roosevelt (VCN-71) and, once in theater, began conducting missions in support of Operation Enduring Freedom.

Where did they begin, these highly-trained individuals known today as Navy SEALS, but popularly known as frogmen during NASA's splashdown days?

It's tough to pin down their origins to a day, week or month but after scanning page after page of military histories, we settled for 1943 (and that is a pretty broad starting point) as the year the Navy's efforts at building Underwater Demolition Units began in both the Atlantic and Pacific.

It is clear that the initial efforts unfolded at a base in Fort Pierce, Florida, which is where the UDT-SEAL Museum stands today. The very first units were known as NCDU's for Naval Combat Demolition Units but UDT's or Underwater Demolition Teams came right along with them and ultimately the acronym NCDU disappeared. The need for this sort of fighting man was obvious to Allied military leaders of that time. Reconnaissance and demolition missions had to be carried out expeditiously along enemy-held beaches in both the Atlantic and Pacific theaters of operation. Some of the earliest such operations appear to have taken place along the beaches of France in preparation for D-Day.

By the time UDT-14 got together in Hawaii they would acknowledge that Fort Pierce, Florida, had gotten started somewhat sooner than they did and that training courses on the east coast were longer.

Despite the fact that these units originated in Florida and Hawaii, neither state is home base to them today. What is now known as SEAL TEAM Two calls Little Creek, Virginia, its home. The U.S. Navy Amphibious Base at Little Creek is part of the sprawling U.S. Naval Base at Norfolk. SEAL TEAM One can be found at the Navy's Amphibious Base in Coronado, California, and is about as close to the equally huge Naval Base in San Diego as Little Creek is to Norfolk.

As the nation's space program got underway, UDT members learned that they could apply to take part in that effort which would be considerably more glamorous and definitely more publicized than their normal missions and nowhere near as dangerous. We, in the news media, always called them frogmen, seldom if ever using the term UDT; and never called them SEALS because a SEAL was another step up in military skills acquired. A UDT could volunteer for that additional training and many of them did. Even after Washington decided that the teams on both coasts would be known as SEALS in January of 1962, there continued to be UDT's operating at both bases and these were the men we came in contact with aboard the recovery carriers. The UDT's weren't even officially decommissioned until the late 1980's.

HOW MANY REHEARSALS ARE ENOUGH?

This is one of those statistics that is impossible to pin down. It is clearly evident that every ship set its own schedule, and if there was any common thread it would have been – practice, practice, practice.

Frogmen began leaping into the Atlantic or into coves and bays in and around Norfolk (Little Creek) well before Alan Shepard took his short sub-orbital ride in 1961.

Virginia. It pointed out clearly just how early the training and preparation had gotten underway. The officer who took the picture pointed out that such activity went virtually unnoticed and unrecognized. He was right.

Robert A. "Pete" Petersen (Capt. USNR Ret) was Lieutenant Junior Grade Petersen when his involvement in the space program began. His earliest leaps into salt water began at Little Creek, where they were given a boilerplate Mercury capsule and several of the new flotation collars. Petersen and five others were later packed off to Cape Canaveral for further training early in the Mercury phase of the program and ended up working with real live astronauts – Air Force Major L.Gordon Cooper and Navy Commander Alan Shepard. Not just real astronauts but actual spacecraft as opposed to the boilerplate models which so many Navy ships were rapidly being equipped with for their own rehearsal regimens.

Petersen is a walking, talking testament to the degree to which NASA and the Navy went to ensure that all elements of a potential Naval recovery force, in either the Atlantic or the Pacific, would be prepared for any eventuality; and, indeed, there would be a few. Bob recalls testing recovery gear from on board oilers, transports, cargo ships, destroyers and other types of ships in addition

Here is Petersen's pictorial proof. A nice NASA photograph taken during one of those "egress" rehearsals off the Cape in April 1963. Bob is at the extreme left while astronaut Alan Shepard is the middleman with his back to us. The other two gentlemen are unidentified. And that is a real Mercury capsule...not a boilerplate "dummy." (Courtesy of NASA)

It was a time even before the flotation collar was ushered into the routine and before the UDTs would get into the act of real recoveries. The point being – they all knew what was coming and they worked long and hard to be ready when their time came to take part in the program.

A photograph taken back in the summer of 1959, almost two years before the Shepard flight, showed the original seven astronauts with a sizeable group of frogmen, all in swim trunks, with swim fins and masks, pausing during a SCUBA training session at Little Creek,

to the carriers. Since a number of spacecraft did descend miles and miles away from those prime recovery carriers, this preparation proved to be both timely and necessary.

In the course of their training routine, Petersen and his teammates worked hard to get the helicopter pilots as close to the ocean as they possibly could. The pilots had a real aversion to working too close to the water because the salty spray kicked up by their rotor blades could do bad things to their aircraft. The frogmen, on the other hand, were wary of jumping out of those helicopters at anything above 10 or 15 feet because they were loaded down with a lot of gear, including those heavy scuba tanks, the flotation collars and rafts packed into zipper bags. Bob says, at just about any height above the ocean, those tanks alone could do nasty things to a man's back when he hit the water.

Petersen and his pals missed their one chance at taking part in a manned recovery when they sailed from Boston aboard the USS Essex for the Mercury flight of Gordon Cooper in mid-May of 1963. It was a long shot to begin with since Essex was the safety valve in case Cooper's flight had to come down in the Atlantic instead of the Pacific, as planned. Cooper did splash down in the Pacific on schedule where the carrier Kearsarge did its job, but Petersen and the rest of his UDT crew had to be dressed (wet suits) and ready, on board the helicopters, blades turning, just in case. Such would be the procedure on all missions.

Proving again that recovery procedures didn't simply appear out of nowhere and work to perfection without effort, Petersen and a fellow UDT member, Dave Smith, would move on to further testing of recovery gear for the Gemini and Apollo projects. They were sent to NASA, Houston, where they worked with helicopter pilots at Ellington Air Force Base where they were still trying to get those pilots closer to the water. As his part in the space program drew to a close, Petersen said that even though he never had the opportunity to make an actual space flight recovery, the experience was one that he would never trade.

Around the time when Pete Petersen was stepping away from work within the space program, then Lieutenant Junior Grade Chris Bent was literally getting his feet wet. Bent, writing in the UDT-SEAL Association periodical "The Blast" back in 2001, saw his path to NASA being defined by the executive officer of UDT-21, Pat Badger, who informed Chris that he would be the OIC (Officer In Charge) of the Gemini 6 and 7 recoveries which would fly within days of each other in order to carry out the first rendezvous of two spacecraft in orbit.

Chris got to work picking two teams of three frogmen each, and he named Ensign Denny Bowman as his assistant

and in charge of the second team, which was completed with the addition of Roger Bates and John Kennedy. Bent added Dave Sutherland and Danny Fraser and, like Bob "Pete" Petersen and his colleagues, they set out to test the waters of Little Creek, Virginia, the flotation collars and the rafts. I like the way Chris Bent described their rehearsal routine. "Practice made perfect," said Chris, "and we did, we did and did." When I covered the recoveries of Gemini 9, 10 and 11, all in 1966, I saw first hand what Chris was referring to…. again and again and again.

Here are the UDT-21 members who worked those two closely related Gemini missions and it was taken aboard the recovery champion – USS Wasp CVS-18. Kneeling in front are Dave "Red" Sutherland at the left with Danny Fraser to the right. The bag between them holds the flotation collar. Back row, from left to right. Gene Warta, Al Baldwin, Jack Kennedy, Joe Hulse, Roger Bates, LTJG Chris Bent - the OIC, and Denny Bowman who Chris selected to head up Team Two. (Courtesy of US Navy)

Warta, Baldwin and Hulse were a third team and their assignment was to recover the R&R (radar) section of the Gemini 6 spacecraft, which they did.

Chris was just the latest in a long line of UDT men continually pushing to get the helicopter pilots as close to the water as possible, coupled with continual practice on packing and deploying the flotation collar and the support raft, then flybys and drops from the helos.

As for his role, Bent says he would be first man out of the helicopter, swimming immediately to the spacecraft where he would connect a hand phone inside the top end of the capsule's RCS section to talk to the astronauts and make certain that they were okay.

Right behind him would be Dave Sutherland and Dan Fraser whose job it was to swim the flotation collar over to the spacecraft, connect the collar harness beneath the capsule and connect the horseshoe shaped ends. At that point they would activate one of the two CO_2 cartridges, which inflated the collar and made the spacecraft a stable platform.

When it came to the celebrated leap out of the helicopter, Denny Bowman points out the sort of problem anyone not familiar with their world would ever think of. When the first two frogmen made that jump and took the scuba gear, collar and raft with them, the helicopter would suddenly become about a thousand pounds lighter so that by the time the last swimmer exited the aircraft, it was no longer as close to the ocean as it had been. Bowman credits the helo crews with working hard to compensate for that sudden weight loss.

They wore their wet suits, adds Bowman, not because of low water

Gemini Pair Prepare For Space Walk Today

Portlander On 'Reception Committee'

Lt. (j.g.) Dennis Bowman of Portland adjusts a flipper as he and fellow Navy swimmers Daniel Fraser, center, and Roger Bates prepare for splashdown of Gemini 9. Bowman and his team will be dropped into water near spacecraft to attach a flotation collar and to aid the astronauts. They are stationed aboard the carrier Wasp. (AP)

This was how the front page of the Portland newspaper looked back in 1966 when Bowman, Fraser and Bates were prepping for the return to Earth of Gemini 9.

temperatures but for protection against the possible belching of Nitrogen Tetroxide from a spacecraft's attitude control thrusters, which could burn the skin. There never were any such incidents, however. Safety first.

When Gemini 6 splashed down on December 17, 1965, both Chris Bent and Denny Bowman were in the Atlantic and in position in their rafts to witness a baffling sight – an aircraft carrier (the Wasp) listing heavily to starboard. The ship had been several hours away at splashdown so the frog teams spent some time relaxing in their rafts as the big carrier approached…. with a distinct tilt. As it turned out it was the direct result of thousands of sailors gathered on the starboard side to see the spacecraft and the frogmen, most of whom had gathered on the capsule's flotation collar by that time. No problem – the Wasp did not tip over.

Denny Bowman probably would lead any list of frogmen as to the number of missions in which he was involved. Bowman was involved in five Gemini's – 6, 9, 10, 11 and 12. I got to know him pretty well as I sailed on the middle three. Not surprisingly for a young man so constantly a part of this glamorous adventure, Denny got a lot of "ink" back in his hometown of Portland, Maine, as well as the Boston papers and a few others.

It was still a new and exciting frontier, this space business, and it did not hurt Denny's stature one bit having been a championship swimmer at local Cheverus High School only a few years earlier. His name was quite familiar even before he became a Navy frogman.

We cannot overlook the contribution to astronaut recoveries contributed by fixed-wing aircraft from both the Navy and the Air Force.

While the earliest evolution of Navy SEALS took form midway through World War II, an outfit known as Naval Patrol and Reconnaissance Wing Five was established back in 1937 and earned its wings providing air security to American shipping in the Atlantic in the years leading up to the war and on into that conflict. They could look to Norfolk as their main base of operations and the familiar Catalina PBY was one of their workhorses.

In For A Dunking

Lt. (j.g.) Dennis W. Bowman and his wife Phyllis relax here with children Teresa Kathleen and Paul Anthony for a few days before Bowman goes away to pluck more astronauts from the water. He's leader of the frogman team assigned to recover the Gemini 12 astronauts next week. (By Staff Photographer Elwell)

Above: As Denny was getting ready for his fifth and last splashdown, the Portland paper caught this shot of the Bowman family with their newest addition, Paul Anthony, on mom Phyllis's lap. Little Paul debuted during one of Denny's earlier Gemini missions and Walter Cronkite gave out the news during CBS's coverage of that flight. Paul's big sister Teresa Kathleen is on Denny's lap.

Right: Here is Bowman leaping out of a recovery helicopter during one of his five splashdowns. Nice illustration of the ever-present green "locating" dye.

Above: Denny forwarded any number of newspaper stories which ran during his UDT days and one of them identified a Maine artist, Gene Klebe of Pemaquid, who had been with us on Gemini 11, and who provided us with a handy sketched page of frogmen's hand signal codes.

Left: Chris Bent submitted this photo of he and his UDT Team winding up the Gemini 7 recovery on December 18, 1965. Chris is the guy with his arms up, signaling the recovery helicopter to hoist astronaut Frank Borman into the chopper. Chris was the Officer in Charge on this mission. Teammate Dave "Red" Sutherland is in the water and Dan Fraser is in the raft with Chris.

The record shows that a new era in Naval Patrol Operation for Wing Five began in the 1960's with the arrival of the durable Lockheed P3A Orion maritime patrol aircraft, which became celebrated for their surveillance of Soviet ships during the Cuban Missile Crisis. Those Orions had a range which helicopters could not match, and, consequently, they could fly from land bases and participate in the Mercury and Gemini splashdowns in the Atlantic as recovery director aircraft. We knew them as "air bosses" and they would always be thousands of feet higher than all the recovery helicopters and operating as recovery coordinators.

That was the Navy side. Air Force pararescue began, much like the UDT world, in mid 1943 when 21 persons bailed out of a disabled C-46 over an unchartered jungle near the China-Burma border. The crash site was so remote that the only possible means of getting to those people was by parachute drop. A Lieutenant Colonel and two medical corpsmen stepped forward and volunteered for that assignment and the concept of pararescue was born. Among those waiting and hoping in that jungle was news commentator Eric Severeid, later of CBS News renown, who would look back at that ordeal with nothing but high praise for his rescuers.

So what does all of that have to do with the space program? Fast-forward please to mid-March of 1966 and Gemini 8. Astronauts Neil Armstrong and David Scott are in a spacecraft spinning at about one revolution a minute and all plans for an Atlantic splashdown go by the board as they fight to first stabilize their Command Module, and then set their sights on what was basically an emergency landing site in the Pacific.

While they would go on to be picked up in a workman-like manner by the destroyer Mason, the first fellow humans they saw, with the destroyer still hours away, were the pararescue men who flew to them from Naha Air Base on Okinawa. They were so timely in their arrival on scene that they could watch the spacecraft hit the water. Three pararescue men parachuted into the ocean and had flotation equipment attached within 20 minutes.

Frogman Denny Bowman pointed out to me that during every Gemini flight there were these Air Force pararescue teams sitting in a hangar somewhere waiting for just such a situation as faced the men of Gemini 8. Talk about preparation. Talk about back up and readiness.

Denny also noted that his Air Force counterparts had to parachute out of a C-130, probably at an altitude of around a thousand feet with full scuba gear – a task which Bowman described as significantly more difficult and dangerous than making a 15 or even 20-foot leap from a helicopter and then to even land close to the spacecraft. In Bowman's own words, "No small feat. A very difficult assignment." Bowman called them unknown heroes. It was not for lack of effort but we came to realize that we would never find photographs, let alone the names, of all the men who leaped from helicopters to ensure the safe return of our astronauts, but in the five group photos and one list of names appearing in this section, we feel we have found many, if not most, of them.

You saw the UDT-21 gang a few pages back. Now, here are the "dandy dozen" chosen from both UDT-11 and UDT-12 for the APOLLO 11 Recovery. It was the first time two teams were assigned to one recovery and that says a lot about the importance of the first lunar landing.

Kneeling at left in the front is the guy we all came to know as the BIG swimmer, Lt. Clancy Hatleberg. To Clancy's left are Mike Bennett and Tom Holmes. In the back row, left to right are: Charlie Free, Lt.Jg. Robert Rohrbach, Jo Via, Michael Mallory, Lt.Jg. Wes Chesser, John Wolfram, Mitchell Bucklew, Lt. Jg John McLachlan and Terry Muehlenbach who is also credited as the photographer. (Muehlenbach photo)

Hatleberg, Chesser, Mallory and Wolfram all received Navy Commendation Medals for their participation in Apollo 11.

APOLLO 12. No photo but a list of names of the UDT-13 men who did that recovery. Dennis Ranalla, Steve Jewett, Eddie Felton, Arles Nash, W.R. Pozzi, James Cousins, Ernie Jahnke and Robbie Robertson.

Now here are the men of UDT-11 again, but it is APOLLO 14, a year and a half after the Apollo 11 mission, and only two members of the team, Lt.Jg. Robert Rohrbach and Mike Bennett were back for this one.

UDT FROGMEN OF THE

UNDERWATER DEMOLITION TEAM II
NAVAL AMPHIBIOUS BASE
CORONADO, CALIFORNIA 92155

APOLLO 14 Recovery Team

EMI M.P.L BENNETT

OFFICER IN CHARGE
DECONTAMINATION SWIMMER
LTJG R.R.ROHRBACH

HMI T.G. "DOC" HOLMES

LTJG F.W. SCHMIDT

LTJG M.L. SLAGLE

EN3 J.E. RIDDLE

YN3 R. DAVIS

SN L. FALLER

RMSN J.J. MC FARLAND

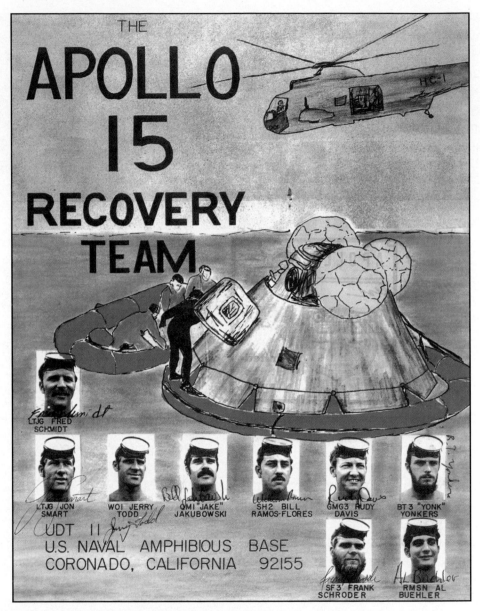

THE APOLLO 15 RECOVERY TEAM

LTJG FRED SCHMIDT

LTJG JON SMART

WOI JERRY TODD

OMI "JAKE" JAKUBOWSKI

SH2 BILL RAMOS-FLORES

GMG3 RUDY DAVIS

BT3 "YONK" YONKERS

SF3 FRANK SCHRODER

RMSN AL BUEHLER

UDT 11
U.S. NAVAL AMPHIBIOUS BASE
CORONADO, CALIFORNIA 92155

Here is another nice unit-created photo page of the UDT-11 members who carried out the Apollo 15 together with helicopter squadron HC-1.

The RECOVERY helicopter (that is. the helo assigned to pick up the astronauts themselves) was piloted by Commander Stephen A. Coakley and co-piloted by Lt. Jg. John M. Murphy, Jr. Crew men aboard were Aviation Machinist Mate First Class Ernest L. Skeen, Aviation Electronic

Technician Second Class Thomas R. Hardenbergh and NASA Flight Surgeon Dr. Clarence A. Jernigan.

The SWIM TWO helicopter was further described as the prime flotation collar helicopter and that meant frogmen. It was piloted by Lt. Commander David D. Cameron and co-piloted by Lt. Jg. Stephen M. Lind. Crewmen aboard were Aviation Machinist Mate Second Class John H. Driscoll and Aviation Electronics Technician Third Class Bryce E. Devonport. The swim team leader of UDT-11 was Lt. Jg. Fred W. Schmidt, swimmer number two was Quartermaster First Class William C. "Jake" Jakubowski and swimmer number three was Yeoman Third Class Rudy R. Davis.

SWIM ONE was piloted by Lieutenant Donald M. Larsen and co-piloted by Lt. Jg. Eric J. Challain. Crew men were Aviation Machinist Mate, Second Class Larry G. Parker and Aviation Anti-Submarine Warfare Technician Airman Thomas F. Sharafik. The swim team leader was Warrant Officer Jerry L. Todd, swimmer number two Ship Fitter Third Class Frank S. Schroeder and swimmer number three was Radioman Seaman Roy Alan Buehler.

By April of 1972 the UDT assignment fell to Team 12 and a dozen fresh faces for the recovery of Apollo 16 (pictured at right).

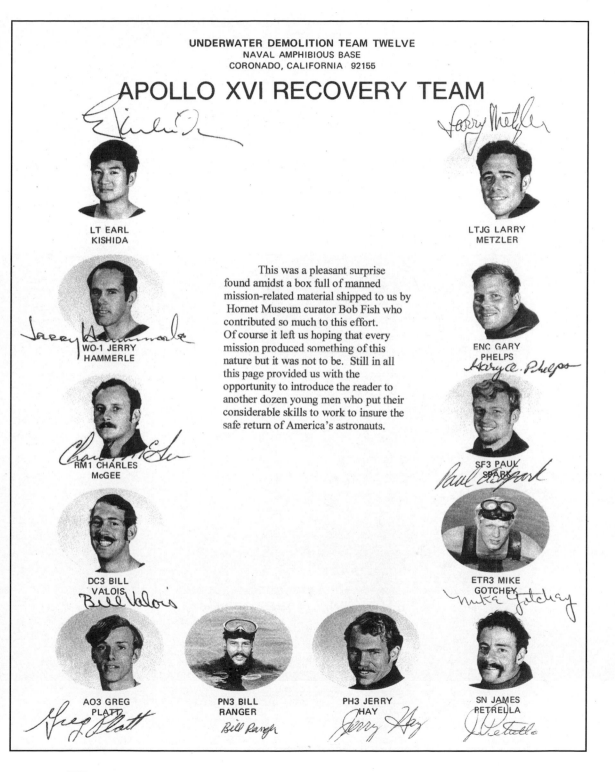

UNDERWATER DEMOLITION TEAM TWELVE
NAVAL AMPHIBIOUS BASE
CORONADO, CALIFORNIA 92155

APOLLO XVI RECOVERY TEAM

LT EARL KISHIDA

LTJG LARRY METZLER

WO-1 JERRY HAMMERLE

ENC GARY PHELPS

RM1 CHARLES McGEE

SF3 PAUL SPARK

DC3 BILL VALOIS

ETR3 MIKE GOTCHEY

AO3 GREG PLATT

PN3 BILL RANGER

PH3 JERRY HAY

SN JAMES PETRELLA

This was a pleasant surprise found amidst a box full of manned mission-related material shipped to us by Hornet Museum curator Bob Fish who contributed so much to this effort. Of course it left us hoping that every mission produced something of this nature but it was not to be. Still in all this page provided us with the opportunity to introduce the reader to another dozen young men who put their considerable skills to work to insure the safe return of America's astronauts.

FROM SIMULATED SPACECRAFT RECOVERY TO THE REAL THING
PRACTICE MAKES PERFECT

Let's take it from the top. A pictorial look at the various stages of a recovery at sea, utilizing both rehearsal photography (primarily my own photos) and actual recoveries (NASA or Navy photos) in which we will skip from one mission to another in order to present a sequential look at the process and the men and equipment involved.

I will include some of the sidebar events which made up our time at sea. First off have a look at a great aircraft carrier whose wartime or post-war record is detailed a little later on in this book, as are all the recovery ships.

The USS Wasp (CVS-18). We boarded her at the oddly named Boston Army Base in early June of 1966 for the flight of Gemini 9 with astronauts Tom Stafford and Gene Cernan. If you looked around for what you expected an Army base to look like, you would be disappointed. In all likelihood the name came from that area's use as a regular port of embarkation for troops but it looked like just another dock, pier or whatever when we got there.

This is a stock photo of the Wasp, more in keeping with her appearance as she performed her anti-submarine warfare role many years after her noble service in World War Two.

USS Wasp, CVS-18. (Courtesy of Former Wasp skipper Gordon Hartley, USN, Ret.)

We sailed, as recovery ships always did, well in advance of an actual lift-off at the Cape. Unfortunately the Agena target vehicle, which Gemini 9 was supposed to rendezvous with, blew up in flight so the astronauts had to be put on hold while NASA scrambled to find the Agena's replacement and get it up there. In the meantime, Wasp was directed to head back to Boston and wait for the next green light.

Grumman S2E (Navy photo)

Those of us not burdened by an excess of equipment were notified to pack and be ready to be catapulted off the ship, back to Bermuda, and from there, via Eastern Airlines (remember them?), back to where we had come from. And this was our air transportation – a compact, powerful Grumman S2E Sky tracker.

They called them COD flights for "Carrier On-board Delivery" and they carried perhaps 8 or 10 passengers and a lot of paraphernalia, often from ship to ship.

We were seated facing the rear of the aircraft and told to lean into our shoulder harnesses anticipating the jolt of the catapult. Nobody said anything about our legs; and when that catapult shot us off the deck, my ankles slammed into the seat in front of me and I wore the bruises for a couple of weeks. Despite that, it was still a great experience for any landlubber.

Just days after we headed out into the Atlantic again to take up a position in the planned splashdown area, an oiler (the Hassayampa) came alongside and both skippers adjusted their speed and distance

apart to allow deployment of the hoses which then brought us all we needed of the various fuels used by the ship and its aircraft. It is a ticklish maneuver and very interesting to watch. This operation, as I recall, took several hours. (See next page.)

The Wasp's captain, Gordon Hartley, and I got along well from day one since I took to hanging around the bridge when the skipper was there, and he had no objection to my presence or my constant questions. Junior officers probably thought Gordon was losing it when he gave up his Captain's chair and told me to "take over." Strictly for laughs. On the day the oiler hove into view he had asked me if I would care to take a ride over to the tanker, have lunch with its captain, swap some movies and maybe bring back some popcorn. I did it all, but only

On any given day aboard Wasp you might find the mid-deck aircraft elevator lowered down to the hangar bay where it made a great venue for volleyball. Plenty of daylight and practically no wind. In this shot you have a mix of Navy, NASA and perhaps a few of us from the news media. (Don Blair Photo)

after he assured me that the sailors manning the ropes on both ships would not get smart and try to dunk me in the Atlantic. The skipper told me that the rope gangs had once dunked an Admiral, but they blamed it on "wave action." They kept me pretty high and dry going over and coming back as you see in the photo.

A few days later (same mission) I took the rope ride over to a destroyer. That ship, the USS Wilson, was giving those who made the trip a nice certificate to verify the episode. The Order of the Salty Highliners it was called and, in this case, signed by the destroyer's skipper with a rather famous name, Commander C. E. Nimitz. Keeping it in the family. The Wilson, by the way, was a plane-guard destroyer on this mission and it shadowed the carrier in case any of its aircraft ever had to ditch before reaching the flight deck or missing it. Handy to have around.

Left: Oiler and carrier side by side. (Don Blair photo)

Right: Author Don Blair on the highline, and his Order of the Salty Highliners certificate (far right), signed by Commander Nimitz. (Don Blair photos)

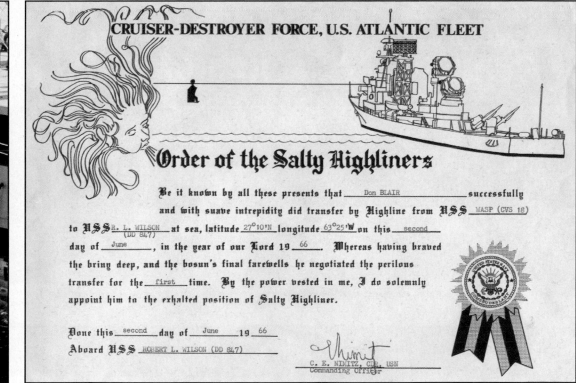

CRUISER-DESTROYER FORCE, U.S. ATLANTIC FLEET

Order of the Salty Highliners

Be it known by all these presents that ___Don BLAIR___ successfully and with suave intrepidity did transfer by Highline from USS ___WASP (CVS 18)___ to USS ___R. L. WILSON (DD 847)___ at sea, latitude ___27°10'N___ longitude ___63°25'W___ on this ___second___ day of ___June___, in the year of our Lord 19___66___. Whereas having braved the briny deep, and the bosun's final farewells he negotiated the perilous transfer for the ___first___ time. By the power vested in me, I do solemnly appoint him to the exhalted position of Salty Highliner.

Done this ___second___ day of ___June___ 19___66___
Aboard USS ___ROBERT L. WILSON (DD 847)___

C. E. NIMITZ, CDR, USN
Commanding Officer

Here is a perfect illustration of the degree of preparedness, which NASA passed along to dozens and dozens of carriers, destroyers, oilers and other ships during the so-called "splashdown" period from Mercury through Apollo and Skylab.

As you can see, a ship would use one of its cranes to lower and retrieve the dummy or boilerplate spacecraft it was given, and all these ships carried out countless rehearsals on their own. As the program would quickly illustrate, those "other" ships would be ships of opportunity on a number of occasions.

Highlining to another ship was always fun and educational but no diversion or opportunity was more exciting or worthwhile than when the ship and/or NASA asked us if we wanted to fly with the frog teams aboard a helicopter on one of those frequent recovery rehearsals – the SIMEX's. I never said no.

So this is how we looked after being issued a nice orange jumpsuit and, armed with camera, awaiting take-off. But the really important guys were headed for a nearby helicopter. (Don Blair photo)

(Don Blair photo)

A typical shot of the UDT guys heading for their helicopter, (frogmen as the world was calling them then) looking deceptively normal but in reality among the most highly-trained and skilled men in any military anywhere. Not your typical men in uniform and they were quietly proud of that. They liked to say that they could do anything and everything land-bound Rangers could do but also do it at sea and that included jumping out of planes or helicopters. UDT stood for Underwater Demolition Team, and in their normal scope of operations they were the best at putting pesky mines on the hulls of enemy ships. For all of these men, the recovery splashdowns were a unique and generally welcome change of pace, and I did not meet one of them who was not pleased and proud to be a part of the nation's space program.

And this was our bird's eye view, over the hoist operator's shoulder at a point where the frogmen had dropped into the ocean, attached and inflated the flotation collar and two rafts. This shot happened to be from a Gemini 11 rehearsal in September, 1966. The camera was a not-very-expensive Kodak Instamatic doing an excellent job.

(Don Blair photo)

Now here is "the real thing"- Gemini 9, June 6, 1966. In a photo taken from the ocean by one of his UDT teammates, you see frogman Roger Bates standing on the flotation collar and leaning in to talk with astronaut Tom Stafford while crewmate Gene Cernan turns and smiles for the camera – a Nykonos given to one UDT member on every splashdown.

Great shot of recovery helo #57 as well and take note of the green dye in the forefront. Since this was my first recovery, I was somewhere up on the Wasp doing a voice-over description of this scene for the television audience. NBC's Bill Ryan had opened our coverage and I was probably wrapping it up at this point.

(Navy photo)

So you now have an idea of what the rehearsals looked like as they unfolded and perhaps a better perspective of a real recovery and its immediate aftermath. What awaited me when I got home was, I think, also quite interesting. (See news column below.)

(Don Blair photo)

This was the scene in the next day or so on the Wasp's hangar bay and the spacecraft was up on a dolly as McDonnell-Douglas (the company that built the Mercury and Gemini spacecraft) engineers swarmed all over it and removed and packed critical equipment.

Somewhere in that time-frame of a few days after splashdown, I managed to sit myself down on that dolly and have a friend take this shot of me which gives you a vivid picture of the spacecraft heat shield and what the sordid heat of re-entry did to it.

(Don Blair photo)

DAILY NEWS, TUESDAY, JUNE 7, 1966

TV *What's On?* RADIO

Gemini Spacecraft Lands In Lap of Televiewers

By KAY GARDELLA

"The whole thing is beyond our comprehension," commented pool correspondent Don Blair aboard the carrier Wasp as he watched the close-to-target landing of Gemini 9, an event destined to take its place in TV history with the first space walk films.

It was the first time televiewers saw a landing of a spacecraft, and to say it was exciting would be the understatement of the day. The pickup from the carrier Wasp, as seen on NBC, CBS and ABC yesterday morning, was a perfect climax to television's excellent coverage of the 72-hour space operation.

Watching the striped, billowing parachute and descending capsule float down into the angry Atlantic, less than two miles off target, and being a witness to the step-by-step recovery maneuvers of Astronauts Cernan and Stafford, was a visual thrill unsurpassed by any sea adventure film.

Adventure at Sea

This was the real thing—trained frogmen making their way in inflated rafts, attaching the flotation collar to the Gemini capsule, swimming out for the life line extending from the Wasp while the ocean threatened their every movement. It was a magnificent live sea adventure with all of the elements of real drama.

To report it for the TV pool was the aforementioned Blair and NBC's Bill Ryan, who turned in a remarkably clear, detailed reporting job. It's always a pleasure to be given a legitimate opportunity to praise this fine reporter, who won the respect of everyone in his profession during the 1962 newspaper strike.

The piece de resistance yesterday, after the descent, was the

Chet Huntley

Bill Ryan

lifting of the Gemini capsule aboard the Wasp and the emergence of the astronauts from their hatches. Momentarily, when the ship's band struck up "Anchors Aweigh," we were reminded of an old Dick Powell movie. Also, the spacecraft, in a TV closeup, looked like an immense Inca mask from the past instead of an important link to our future.

The picture signals were sent to the Early Bird satellite and then to the United States and picked up by the networks.

Added Difficulty

Chet Huntley, in top form, observed that a new obstacle was introduced in this landing. Besides the usual threats of wind velocity, drift, etc., "they now have to clear our antenna structure," he commented wryly. At another point, when the astronauts' meager expense account was discussed, Huntley quipped: "Roy Neal is down in Houston on

a somewhat bigger expense account than that."

The Huntley-Brinkley news team consistently does such a good job, it's unfortunate their desk has to look like a Gulf filling station. As the sponsor for NBC's coverage, the petro outfit gets enough mileage out of its money without this abrasive touch. Unfortunately it negates to an extent, the firm's excellent commercials, which we've pointed out before, are exceptionally good.

Improved Coverage

Looking back over the many space programs telecast, it seems to us the networks have come a long way in their simulated coverage, presenting remarkable animated illustrations that are frequently easier to follow than the actual maneuver or place of equipment being explained. The re-entry, for instance, was shown three different ways by NBC, ABC and CBS, but all visually clear.

Yesterday's ceremonies aboard the Wasp and the astronauts' departure for Cape Kennedy were televised by all three networks. NBC beamed the proceedings overseas via Early Bird at 3:30 P.M.

During these afternoon festivities, a sense of relief was felt as the astronauts cut the elaborate cakes baked in their honor. A humorous and touching moment was struck by Seaman Rogers, who, while presenting Bosun stripes to our space heroes, choked up and became almost inaudible.

Difficult to understand, too, was President Johnson, who was heard in a pre-recorded telephone conversation congratulating Cernan and Stafford.

Lost Contact

For a brief moment yesterday NBC lost the Wasp picture while CBS and ABC remained in contact. Kiddingly, NBC's Frank McGee pleaded: "Come on, Bill. Come on, Ryan. Speak to us."

NBC, which was ahead from the start in the ratings, garnered a 54% share of audience yesterday during the 9 to 11 A.M. period of common coverage.

I hope that Kay Gardella of the New York Daily News realized what I had been trying to say. That any technology such as we had just been eye-witness to was well beyond the understanding of mere mortals – the news media included. By the time of the Gemini 9 flight, Bill Ryan of NBC had quite a few splashdowns under his reportial belt and any praise which came his way was well-deserved.

But the icing-on-the-splashdown cake for me was a note sent to WHNB-TV in West Hartford, Connecticut, where I had been working on-camera in news, weather and sports as recently as the previous December. The letter was forwarded to me at Mutual in New York. It's a great little note. A wife trying to prove she knows more than her husband – just this once.

USS Guadalcanal (Courtesy US Navy)

Now then, let us move on to other missions, other ships and other astronauts.

Not all the carriers used in the splashdown era were Essex or Ticonderoga class. You'll get an explanation of what the differences were between those two later on, but there was another class of carrier that got into the program and this was the first – the USS Guadalcanal – an Iwo Jima class LPH. That stood for Landing Platform Helicopter and told all who cared to inquire that it was designed for helicopters only. No fixed-wing, prop or jet aircraft coming down their decks....no arresting cables....and a noticeably different rounded bow, not to mention being several hundred feet shorter. The main mission of an LPH – to get several thousand combat-ready Marines ashore somewhere in a hurry. But not now. When Guadalcanal got the word from the Chief of Naval Operations, someone in obvious authority directed a lot of her crew to get out on the deck and spell out Gemini 10.

(Don Blair photo)

(Don Blair Photo)

Terry Drinkwater and Don Blair on the deck of the USS Guam. (Courtesy of NASA)

What you are looking at are two astronauts who had a lot more history to make following this mission. Closest to us is John Young who would go on to fly around the Moon in Apollo 10, actually walk on the Moon during Apollo 16 and be the first Shuttle pilot in 1981. An enviable, admirable record. Facing us is Mike Collins who minded the store in the Apollo 11 Command Module while Neil Armstrong and Buzz Aldrin became the first two men to ever get right down and walk on the lunar surface. But what is that youngster doing in this picture? It was summertime in Norfolk and the Navy decided to allow a few officer's kids to make the trip of their lives. I don't recall exactly how many we had with us but there were only a few – a very lucky few.

On to Gemini 11 in the fall of 1966 and the second LPH to take part in a recovery, the USS Guam (LPH-9), and a carbon copy of her sister ships in almost every instance.

Here is good look at the deck of the Guam as the late Terry Drinkwater of CBS (at left) and I were checking out broadcast positions and a NASA photographer, with not much else to do until the splashdown, followed us around one morning and got this shot. I'll never forget how Terry worried about the NBC technical crew on board. It seems they got their gear up and running in all corners of the ship in only a few days and Terry was sure they must have overlooked something. They hadn't.

Over the course of five splashdown assignments I never found Navy officials anything but pleasant to deal with. They were, in almost all cases, approachable and generous with their time although we did have to ask for interviews in advance.

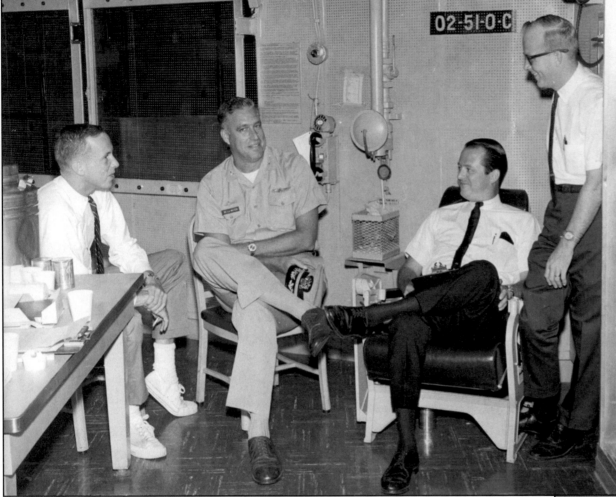

This is the Guam's skipper, Captain Steve DeLaMater, putting up with our questions and curiosity one day. Terry Drinkwater of CBS is seated left, then the Captain, myself and Mike Clark of United Press (standing) at right. (NASA or Navy photo)

This photo shows a typical flight deck with recovery helicopters lined up and ready to go and the huge TV dish down near the stern all tuned up to give the folks at home a first hand view on splashdown day. (Don Blair photo)

Now here are a few Guam publications issued just before the big day – splashdown. I offer them without comment because after 37 years there is no way to learn the full names of any individuals mentioned as on the September 15 flight schedule for instance. They will, however, offer a glimpse of the composition of the recovery day air group (all helicopters), their designations (as in Search 1, etc.) and their pilots and co-pilots.

THURSDAY, 15 SEPTEMBER 1966
FLIGHT SCHEDULE

EVENT	AIRCRAFT	PILOTS	LAUNCH/RECOVER	MISSION	FUEL	CALL SIGN	FLIGHT QTRS TIME
2	3 SH-3A	MILLER/SMITH	0916/COMPL	GEMINI XI RECOVERY	4+30	SEARCH 1	0815
		TRYGSLAND/SPOFFORD	" "	"	"	SEARCH 2	
		DAWSON/NAVONE	" "	"	"	SEARCH 3	
3	3 SH-3A	DOEGE/ROTSCH	0941/COMPL	GEMINI XI RECOVERY	4+30	SWIM 1	0815
		HORNING/LEWIS	" "	"	"	SWIM 2	
		KURATH/GOODING	" "	"	"	SWIM 3	
1	1 SH-3A	HS-3	0630/----	LOCAL TEST	4+30	TROUBLESHOOTER #58	0600

W. G. Davis
W. G. DAVIS
OPERATIONS OFFICER

USS GUAM LPH-9
GREEN SHEET

USS GUAM · LPH · 9

SUNRISE: 0625Q
SUNSET: 1852Q

THURSDAY, 15 SEPTEMBER 1966
(RECOVERY DAY)

ALL TIMES QUEBEC AND ZULU

TIME QUEBEC/ZULU	EXERCISE	REMARKS/NOTES
	SYNOPTIC & SPECIAL WX. REPORT	
0200/0600	SPECIAL HOURLY WX. REPORT	
0400/0800	SPECIAL HOURLY WX. REPORT	
0500/0900	FLIGHT QUARTERS	
0600/1000	SPECIAL HOURLY WX. REPORT	
0600/1000	GUAM, ON STATION (RECOVERY AIM POINT)	LAT. 24-16N LONG. 78-00'W SUBMIT ON STATION REPORT TO CTG 140.3
0600/1000		
0630/1000	LAUNCH HELO IAW FLIGHT SCHEDULE CTG 140.3 SUBMITS ON STATION REPORT	
----/----	SECURE FLIGHT QUARTERS	
0700/1100	SPECIAL HOURLY WX. REPORT	
0700/1100	METRO, LAUNCH WX. BALLOON	
0715/1115	GUAM SUBMIT FUEL AND POSITION REPORTS	BRIEFING OFFICERS: MR STONESIFER (NASA) CDR. DWYER, LT. SNELL
0800/1200	PRESS CONFERENCE IN SACC (WX. & OPERATIONS BRIEF)	
0800/1200	SPECIAL SYNOPTIC & RAWIN WX. REPORT (HOURLY)	DROGUE - 50,000 FEET REEFED
0800/1200	MAN ALL GEMINI XI FINAL RECOVERY STATIONS	
0830/1230	TIME CHECK (INITIATED BY GTG 140.3)	To Pilot - REEFED
0830/1230	FLIGHT QUARTERS	R & R CAN OFF + MAIN OUT AT 10,000
0815/1215	SPECIAL HOURLY WX. REPORT	REEFED
0900/1300	LAUNCH SEARCH 1, 2 & 3 HELO's	FULL DEPLOY AT 9,000
0916/1316	RETROFIRE	Two POINT AT 6,000
0925/1325		

TIME QUEBEC/ZULU	EXERCISE	REMARKS/NOTES
0941/1341	LAUNCH SWIM 1 & 2 AND PHOTO 1	
0941/1341	AIR BOSS 1 & 2 ON STATION CTG 140.3 SUBMIT A/C ON STATION REPORT	
0949/1349	BLACKOUT, (COMMUNICATIONS) BEGINS	
0954/1354	BLACKOUT EXPECTED TO TERMINATE	
0955/1355	DROGUE CHUTE DEPLOY 50,000 ft. PILOT CHUTE DEPLOY 10,600 ft. R AND R SECTION JETTISON AND MAIN CHUTE DEPLOYS.	
0956/1356	MAIN CHUTE DIS-REEF, 9,800ft. TWO-POINT SUSPENSION	
1000/1400	SPECIAL HOURLY WX. REPORT	
1001/1401	SCHEDULED, GEMINI XI SPLASHDOWN	
----/----	ASTRONAUTS ARRIVE ABOARD FLIGHT DECK	WELCOME ABOARD CEREMONY
----/----	ASTRONAUTS PROCEED TO SICK BAY	
----/----	LOAD SPACECRAFT ABOARD	DELIVERY OF MATERIAL TO CAPE KENNEDY
----/----	STAR AIRLIFT FROM FLIGHT DECK AFT	
----/----	GUAM, UNDERWAY FOR MAYPORT, FLORIDA AT 20kts. ETE IS 34 HOURS, 30 MINUTES.	
1430/1830	GREEN SHEET CONFERENCE	STAFF WARDROOM
1530/1930	PRESS CONFERENCE IN SACC	BRIEFING OFFICERS: MR STONESIFER (NASA) CDR. DWYER LT. SNELL
1600/2000	PRESENTATION CEREMONY ON HANGAR DECK	ALL HANDS

W. G. Davis
W. G. DAVIS
OPERATIONS OFFICER

The Guam Green Sheet is a nice chronology of events which began unfolding many hours before anticipated Gemini 11 splashdown. You can follow the process through the early briefings, weather reports, flight quarters and then the return to Earth of astronauts Pete Conrad and Dick Gordon. The notes are my own and were supposed to make me sound more intelligent on the air. Hope it worked.

Here is one of the best photos I have seen in all the years since these events took place. There was always one helicopter with NASA and Navy photographers on board and this shot had to come from one of them. They were not always identified as Photo Helos however. Kudos to whoever took this because it shows us astronaut Pete Conrad scrambling out of the Gemini 11 spacecraft, an excellent look at the flotation collar and the UDT raft attached, fine detail of the spacecraft itself and the ever-present and useful green dye out near the nose of the spacecraft. This shot was taken when Gemini 11 was still a few miles from the ship.

Here is a look at the conclusion of the actual Gemini 11 mission as we were broadcasting it. In this first shot the recovery helicopters have touched down, the astronauts are stepping down from that helicopter nearest to us and I am broadcasting the event from my 06 perch. I'm the guy in the blue shirt, hard right, leaning out from the semaphore position.

(Don Blair photo)

Now that we have that taken care of, fast forward a few minutes and the astronauts have arrived mid-deck for a ceremony that was repeated on just about all missions with the exception of those three Apollo quarantine missions still to come. Conrad and Gordon are at the microphone, thanking the ship and its crew for a recovery well done. And why not? They were all well done. Note the red carpet, center piece of a celebrated caper which we describe later in this book.

That did it for me in 1966. Three recoveries in a row and I was starting to feel like an honorary sailor and enjoying every day at sea, but then matters got real quiet throughout 1967 and 1968 for me as the "luck of the draw" at the network news executive meetings in New York

(Navy photo)

(Don Blair photo)

simply did not go our way (the Mutual Radio Network). I began to wonder if I would ever get out on an aircraft carrier again with one mission in particular uppermost in my mind and on my wish list.

As it turned out it was my good fortune to be a part of the expanded correspondent pool assigned for Apollo 11 - the mission so many of us really wanted. Those news directors who had been picking network names out of a hat or ashtray and sending out just two reporters on all previous missions, wisely decided that public anticipation for this first-ever attempt at a Moon landing merited a correspondent from all the networks.

Consequently NBC sent Ron Nessen, CBS sent Dallas Townsend, ABC sent Keith McBee and I got the catbird seat on the 06 level, four levels above the flight deck for the only radio reportage coming from the Hornet on the real splashdown day. The Mutual network, my home base in those days, was a radio-only network throughout its history.

Don Blair at 06 level. (Don Blair photo)

From my very first carrier assignment three years earlier I had gravitated to a little cubicle along the walkway on the Wasp's 06 level which gets its number from the number of flights above the hangar bay....not the flight deck. I was able to keep that spot through five splashdowns and always thought I had the best broadcast spot on any of them. This is a little bulge of an area where a sailor would stand to show his semaphore flags to other ships. It was just outside CIC (Combat Information Center), the plotting room where the Navy and NASA people that I had to hear from to carry on my coverage, were bending over a large table, lit from underneath, and on which all movement related to the ship's current deployment was being displayed. Little model ships, planes or, in this instance, a spacecraft.

While I headed for a higher perch, my broadcast colleagues usually

(Don Blair photo)

And this was our home....the battle-tested USS Hornet. I won't go into her incredible wartime record here. That comes later but it was an awesome record and when we boarded her at Pier Bravo in Pearl Harbor, her hangar bay was still proudly showing all the Rising Sun emblems indicating the huge number of Japanese planes and ships downed during the last year and four months of the war in the Pacific.

operated from down around the flight deck or even in the hangar bay below it. Here is NBC's Ron Nessen looking ready to break out with a good broadcast even before the Hornet had departed Pier Bravo at Pearl Harbor. Ron, of course, went on to become press secretary to President Gerald Ford in the mid-70s.

Soon after we departed Pearl the recovery ship was back in the SIMEX

Ron Nessen with broadcast gear. (Don Blair photo)

(Don Blair photo)

(simulated recovery rehearsal) business and I was back out in one of the recovery helicopters in time to catch the BIG (Biological Isolation Garment) swimmer, Lt. Clancy Hatleberg, in mid-air as he dropped from famed chopper #66 into the Pacific, where his UDT mates had already done a nice job of attaching and inflating the flotation collar and raft. Number 66 would turn in five well-done recovery assignments including, of course, Apollo 11.

Now have a nice close look at what an Apollo dummy spacecraft looked like. This shot was taken from the flight deck with a rather impressive 500mm Canon lens on my F-1 camera body. Those guys were really way out in the Pacific. Great lens. You can see the looped cable used to lower the dummy into an ocean for the rehearsals and also the little antenna that sent out signals the recovery helicopters homed in on. Some ships painted their boilerplate spacecraft. Hornet didn't.

Now here is something which became a common sight after every SIMEX. Flotation collars washed off and drying out on one of Hornet's elevators before being repacked for the next rehearsal.

Apollo boilerplate capsule. (Don Blair photo)

Flotation collar drying on Hornet's elevator. (Don Blair photo)

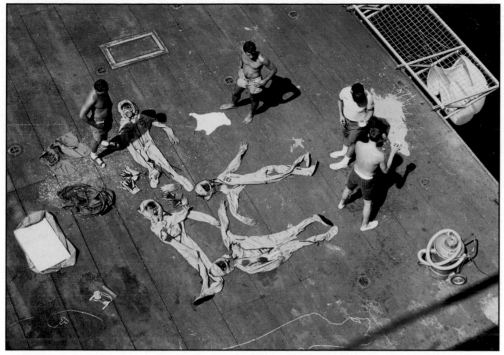

BIG suits drying. (Don Blair photo)

More than a few key people in NASA lost a lot of sleep worrying about what sort of new germs our astronauts might pick up on the lunar surface. Some of their colleagues ridiculed their concern but the worry-warts won out and caution came to fruition in the form of the BIG's.

About those BIG garments. It makes a lot sense when you realize that the frogmen, Lt. Hatleberg certainly included, would wear these lightweight zip-up suits with gas masks on them that would filter anything they might breath in, while the garments which Clancy would toss into the spacecraft to the Apollo 11 crew would have masks designed to filter out anything they would exhale. When the real thing came along, Hatleberg and his colleagues donned their BIG suits, Clancy washed down the Command Module with a surgical wash called Betadine – dispensed from what were basically modified fire extinguishers – and then plugged in an exterior phone to ask the crew if they were okay and ready for their part in this "germ warfare" thing. The spacecraft hatch was then opened, the suits thrown in and the hatch quickly shut until the astronauts again signaled that they were dressed and set to be hoisted up into a helicopter.

But before we would get to that great moment when history's first Moon walkers would get back to our waiting recovery ship, we welcomed our President. Here is Mr. Nixon as he first appeared in the doorway of Marine One, which had just brought him in from near-by and super-secret Johnston Island. The President was beginning an extensive world tour at the time.

(Don Blair photo)

Here is the President as he moved through a sort-of receiving line which included Capt. Carl Seiberlich, Commanding Officer of the Hornet; Dr. Tom Paine, Director of NASA, standing near the President's left shoulder; and Admiral John McCain, just in back of

(Don Blair photo)

Mr. Nixon where he is talking to another officer. Admiral McCain was CINCPAC, Commander-in-Chief-Pacific, and even then his son, John III, was a prisoner of war in Vietnam.

As for the sailors and their colored outfits. They were called Rainbow Sideboys and it did serve to brighten up the occasion but the colors were more serious than that. They designated a sailor's job on the flight deck (fueling, armament, catapult, etc.), one of the most dangerous places in the world.

Move forward a few hours after the President's arrival and to paraphrase Houston's familiar, "We have lift-off," the ship was telling the world, "We have splashdown." It occurred after a few of us had glimpsed Apollo 11 streaking overhead at just over 10,000 feet, through broken clouds and on its way to imminent main parachute deployment and a splashdown just over our horizon some 10 or 11 miles from the Hornet.

Needless to say the now well-tuned, well-trained recovery helicopters and their UDT teams were over the Command Module in mere minutes. Here is another seldom seen photograph showing Lt. Clancy Hatleberg, the BIG swimmer, securing a hatch on the Apollo 11 Command Module while the astronaut crew looks on. Obviously they are all wearing their BIG suits and would continue doing so until the #66 recovery helicopter brought them to one of Hornet's elevators and they strode across the hangar bay and into the Mobile Quarantine Facility (MQF) whose design was based on the 32-foot Airstream travel trailer.

(Navy photo)

My broadcasting ended when the Apollo crew was safely on board and I became the picture-taking tourist again. This picture was one of the results. (Don Blair photo)

On either side of me were professional photographers from *Life*, *Look* and *National Geographic* and any number of others including many from other countries. A week or so later when *Look* brought out an issue devoted entirely to the Apollo 11 mission, their photo was almost identical to mine. After all, we were looking at the same thing at the same moment, all of us snapping pictures like mad..

It did not take long for the astronauts to shed the BIG suits and at least get reasonably cleaned up in order to appear in the MQF window and greet the President of the United States and the world. It was the proper time for Chaplain Piirto's prayer of Thanksgiving. (See next page.)

The world has seen one particular picture of President Nixon, in a slight crouch, leaning toward the forward end of the MQF and talking to the crew of Apollo 11. You can well imagine how many camera shutters were clicking while all this was going on. Shortly after the President bid his good-byes and climbed back into his helicopter, further ceremonies continued including a beautifully decorated cake, more chit-chat between astronauts, NASA and Navy personnel.

USS HORNET (CVS-12)

Date ___24 July 1969___

From: The Chaplain

To: Press Center

Subj: Thanksgiving Prayer, copy of

1. The following Thanksgiving Prayer will be prayed this morning:

Lord God, our heavenly Father, our minds are staggered and our spirits exultant with the magnitude and precision of this entire Apollo 11 mission. We have spent the past week in communal anxiety and hope, as our astronauts sped through the glories and the dangers of the heavens. As we try to understand and analyze the scope of this achievement for human life, our reason is overwhelmed with abounding gratitude and joy, even as we realize the increasing challenges of the future.

This magnificent event illustrates anew what man can accomplish when purpose is firm and intent corporate. A man on the moon was promised in this decade, and though some were unconvinced, the reality is with us this morning in the persons of Astronauts Armstrong, Aldrin and Collins. We applaud their splendid exploits and we pour out our Thanksgiving for their safe return; to us, to their families, to all of mankind. From our inmost beings we sing humble, yet exuberant praise.

May the great effort and commitment seen in this project Apollo, inspire our lives to move similarly in other areas of need. May we, the people, by our enthusiasm and devotion and insight, move to new landings in brotherhood, human concern and mutual respect. May our country, afire with inventive leadership and backed by a committed followership blaze new trails into all areas of humanity cares.

See our enthusiasm, and bless our joy with dedicated purpose towards the many needs at hand. Link us in friendship with peoples throughout the world, as we strive together to better the human condition. Grant us peace, beginning in our own hearts, and a mind atuned with goodwill towards our neighbor. All this we pray, as our Thanksgiving rings out to thee, in the name of our Lord. Amen.

J. Piirto

J. Piirto

(Don Blair photo)

Note the slogan over the window – Hornet + Three. Pretty obvious that safety of the ship's crew and recovery of the astronauts was paramount on this mission.

Here is my distinguished colleague, the late Dallas Townsend of CBS, still at work, wearing his headset and describing the colorful hangar bay ceremonies to a world-wide TV audience.

(Don Blair photo)

(Don Blair photo)

A small group of Hornet's crew had completed their enlistments during this cruise. Here is Apollo 11 astronaut Buzz Aldrin, second man to walk on the Moon, administering the oath of re-enlistment. A special moment for those young men.

This shot (below) has special meaning because it gives me the chance to introduce you to a number of individuals who will be mentioned often in this book starting with Dr. Don Stullken, at the right, and wearing the dark suit. Doc led the team which developed the flotation collars and presided as NASA's top man aboard many manned missions. Next to Doc is Captain Carl Seiberlich, skipper of the Hornet while the gentleman in the dark suit at left is Dr. Tom Paine, the head of NASA. Standing in the foreground with his back to the camera is the late Ben James, NASA Public Information Officer.

(Don Blair photo)

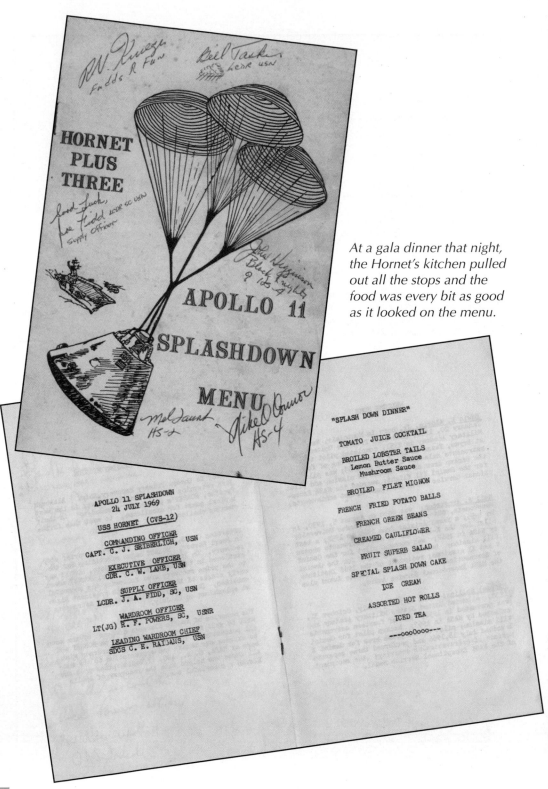

At a gala dinner that night, the Hornet's kitchen pulled out all the stops and the food was every bit as good as it looked on the menu.

While not a photographer by trade, I have been in position to take some very significant photos; and while this next picture probably could have come out sharper and better defined in the hands of one of those old pros on board, I am as proud of this one as any picture I have ever taken. More credit to the camera and the lens than to the guy who pushed the button.

(Don Blair photo)

It was around ten the night of the recovery (July 24). Not having enough sense to retire and get some well-earned rest, I decided on one more trip down to the hangar bay and the MQF just to see "what was up." There was a lone Marine guard to the left of that front-end window where the astronauts had chatted with their President hours before. As I approached I could see one man standing in the living room, and as I moved in as close as the Marine was going to allow, I realized it was Neil Armstrong and that he was strumming a ukelele. I later learned that he carried that little instrument around with him a lot.

I did not have flash equipment. Never expected to need it on an assignment of this nature. Therefore, with the F-1 loaded with Kodak's fast Tri-X black-and-white, I pushed the lens to obliterate the window and popped off three quick shots of Neil and his mini-orchestra. I turned, walked away, looked back a few seconds later, and the first

man to walk on the Moon was gone, maximum photo-op. There and gone and I was the only one who got the shot. The other two were even less sharp than the one you're looking at but I was well pleased with this effort since I still feel the diffusion of the overhead lighting, halo-like, was more than fitting for a man who so recently ventured to near-heavenly heights.

When Neil stepped down from the space program and took up a position teaching aerospace at the University of Cincinnati, I sent him an enlargement of this picture. I did not hear from him, but I didn't get the picture back either, so I assume he got it. Neil was never the poet laureate, golden-tongued after dinner speaker that many followers of our space program might have wished. A few have observed that "the wrong guy got out first" and, indeed, Buzz Aldrin was the more talkative of the two, but for many others Neil Armstrong was just fine…modest and soft-spoken.

Before we get off the ship and broadcast the off-load of the MQF with the three astronauts and two NASA personnel inside, let's have a look around the hangar bay where not just the astronauts were hunkered down in their specially modified Airstream, but NASA friends were busy deactivating Columbia, the Apollo 11 Command Module. There were those priceless Moon rocks to be unloaded and all sorts of equipment to be removed and packed for the trip back to Houston. This is how it looked.

(Don Blair photo)

There was another trailer behind the one occupied by the astronauts. Back-up. NASA's middle-name is back-up. We would often see shadowy figures moving through that tunnel as the astronauts and either the NASA doctor or the NASA technician with them, went back and forth. Note the balloons used to pop Apollo 11 upright in the Pacific are still inflated.

I mentioned the Betadine surgical wash which Lt. Hatleberg used to spray down the spacecraft to kill "Moon germs" if there were any. Here they are arrayed among boxes, tanks and other containers being used to pack away anything and everything from inside the Command Module. Some of the tanks we saw held all the astronaut's "waste."

(Don Blair photo)

During this time of doubt and concern over out-of-this-world bugs, everything the astronauts touched had to be saved and examined.

Here is a picture that will make many camera buffs take stock. A

(Don Blair photo)

Hasselblad camera and a real NASA favorite on just about all manned missions. Worth thousands, this is one of the lucky ones that was not left on the Moon in a trade-off, weight-wise, for Moon rocks. There are quite a few up there for the taking but the cost of a round-trip is rather prohibitive even for frequent flyers.

(Don Blair photo)

Time to go home. We are back at Pier Bravo, Pearl Harbor, and the welcome home festivities are underway as we broadcast the offload. Five men are still in that trailer. Besides the Apollo 11 crew, there was NASA MQF technician John Hirasaki and NASA flight surgeon William Carpentier. The MQF, sitting on that specially built hydraulic-lift trailer (and there was a back-up trailer of course) was later driven over to Hickam Air Force Base where it was loaded aboard an Air Force cargo plane and flown back to Houston.

After all five men had been in the Quarantine trailer for 3 days, they were at least able to exit into a larger room called the LRL – Lunar Receiving Laboratory, where they remained in a quarantine situation and beyond physical contact with the "outside world" for the remainder of the three-week quarantine period.

Before we move on to a few photos we took during our last splashdown assignment, Apollo 15, here is something close to a collector's item sent to us by then Commander Chuck Smiley, the pilot of the celebrated helicopter number 66 during the trouble-plagued Apollo 13 mission. Chuck had previously piloted number 66 during Apollo 10. Here is the flight operations sheet for Apollo 13 on recovery day.

FLIGHT SCHEDULE FOR FRIDAY   SDO: LT CAMPBELL
17 APRIL 1970

EVENT	BRIEF LAU REC	PILOTS/CREW	MISSION	NOTES
1	0545 0700 TBA	*CDR SMILEY/McCARTHY BAIRD/JAHNCKE Slider/Longe	APOLLO 13 RECOVERY	1,2
2	"	*FRANK/HUDSON Petersen/Nielsen Palma/Phanzelter/Jewett	APOLLO 13 RECOVERY SWIM ONE	1,2
3	"	*WILLHITE/WRIGHT Morrison/Towne Starr/Banfield/Carolan	APOLLO 13 RECOVERY SWIM TWO	1,2
4	"	*HIGGINSON/TURNER RHODER/WHALEY Hatfield/West Martin/Howard	APOLLO 13 RECOVERY PHOTO	1,2
5	"	*WALKER/MILLER Dennison/Disco	APOLLO 13 RECOVERY RELAY	1,2
6	"	*CDR AUT/WHITTEN Dominguez/Wardwell	APOLLO 13 RECOVERY RECOVERY/SWIM BACK-UP	1,2
7	1100 1200 (ABT) 1300	*DUNCAN/LYBARGER Broyles/Garren Palma/Banfield	STAR PLANE GUARD	
8 9 10	TBA TBA TBA	*STRAWN/TILLMAN Faurot	DUTY TEST	
11	TBA TBA TBA	*LINKER/LYBARGER Longe/Nielsen	SAR STANDBY	

15 APRIL HOURS FLOWN	16.3
APRIL TO DATE	124.4
TOTAL NIGHT	0.0
APOLLO TO DATE	184.6

FLIGHT NOTES: (*) DENOTES PILOT IN COMMAND
              1. LCDR MILLICAN BRIEF LAUNCH
              2. DEBRIEF TO BE ANNOUNCED

NATOPS STATEMENT: THE CRUISE GUIDE INDICATOR WILL READ DOUBLE IF ONE INDICATOR IS
                   INOPERATIVE.

SUBMITTED: APPROVED:

J. J. HIGGINSON   C. B. SMILEY
OPERATIONS OFFICER   COMMANDING OFFICER

(Courtesy of Chuck Smiley)

And here is the crew. Left to right, Commander Chuck Smiley, his co-pilot Lt. Dana McCarthy, Petty Officer Mike Longe and Chief Petty Officer Glen Slider. Chuck seemed to downplay his own role in the recovery when he wrote that his job was to "simply" put the helo into position directly over the recovery raft attached to the Command Module. Nothing simple about it. The rest of the crew, said Chuck, did the hard work. Dana McCarthy kept an eye on the overall operation and monitored the instrument panel. Glen Slider talked him into position, kept him there and also operated the hoist while Mike Longe (2d crewman) assisted the spacecraft crew out of the Billy Pugh net and got them settled in the helicopter's troop seats. Chuck added that it always looks simple when everything works, but they all breathe collective sighs of relief when their precious passengers are deposited on a recovery carrier's flight deck.

My splashdowns, as it turned out, had one more to go. In typical American fashion, as fast as we demonstrated our expertise and dexterity in a simple little thing like landing men on the Moon, public interest dropped like a rock. It was ho-hum time…the cost of excellence, and so the news directors had gone back to the "two correspondents aboard" routine and my network, Mutual, did not win the draw again until Apollo 15. That was Dave Scott, Jim Irwin and Al Worden and they just kept bringing back more and more Moon rocks but from different locations every time. My broadcast colleague would be David Horwitz of NBC Los Angeles, who would go on to fame (and I assume – fortune) as a hard-hitting consumer reporter who would often turn up on the Today Show in the following years.

This time something occurred that would only happen a couple of times in the entire splashdown era and, of which, I have yet to see any other photographic record.

Quarantine time had become history. The astronauts had been picked up and brought aboard the USS Okinawa the old-fashioned way – without those crazy BIG suits – and would walk down the deck, stand behind a few microphones and thank everybody for a great recovery. But before that could even take place, there we were again on the catwalks that wind around every carrier just below the flight deck level, taking our pictures (some of my most vivid of a spacecraft on the ocean) and suddenly seeing something just below the surface. A few of us yelled, almost simultaneously, "What is it…what is it…and whatever it is…get it!"

What it was, was….the apex cover (some called it a nose cone) of Apollo 15. When the time comes for that device to be blasted away

(Don Blair photo)

Anyway, we had one right in front of us, and I was fortunate to be in position to pop off a few pictures and now we can all take a good look.

In the event that anyone had any question as to which UDT unit did the job on Apollo 15, here is a close-up on the raft they attached to the Command Module with their unit number on it for all to see and their feisty, cigar-chewing frog logo on the floor of the raft. UDT-11.

and the first of the descent drogue chutes deploys, the apex cover is supposed to descend on a small parachute of its own. A number of these covers have been picked up and are said to be stored away in warehouses somewhere but their recovery, obviously, was never as important as the Command Module itself and the men inside it.

(Don Blair photo)

(Don Blair photo)

How about one more Command Module photo for the road? This would turn out to be my last recovery mission and, as I think this picture will attest, produced some pretty vivid close-up shots. Lots of gold-colored kapton foil peeling off the sides of Apollo 15. The CM hatch door malfunctioned and had to be tied shut.

(Don Blair photo)

All ships are in the publishing business to some degree. Some put out more ambitious newsletters than others, but they all do something – whether it's a daily report or a weekly. The USS Hornet put out a hard cover, generously illustrated book on Apollo 11, which I will treasure forever. The Guam published a paperback, but still very respectable product, following Gemini 11 in the fall of 1966. Then there were frequent daily news sheets, such as the one turned out by the USS Okinawa the day following the return of the Apollo 15 astronauts. While the caption says, "Welcome home astronauts Scott, Irwin and Worden," the photo did not agree. From left to right it was Jim Irwin, Al Worden and David Scott. I doubt that it bothered the Apollo 15 trio.

Do you save letters of commendation? And do your friends ever regard them as blatant bragging? I've had a few over the years, but never one that was as appreciated and well-received (by my employers) as the one appearing below. Don Dewsnap was a fellow Mutual Network employee, but in this instance we were both working "pool" for all the networks. This is the kind of letter you save.

DON BLAIR-APOLLO 15 RADIO POOL COVERAGE

TO: CHARLES A. KING
CC DICK ROSSE
VICTOR DIEHM
DUDLEY COX

FROM: DON DEWSNAP

I WISH TO TAKE THIS OPPORTUNITY TO MAKE A FEW COMMENTS ON A FINE RADIO COVERAGE OF THE APOLLO 15 RECOVERY, FROM THE USS OKINAWA. IN MY OPINION, DON BLAIR'S WORK WAS EXTREMELY PROFESSIONAL. I HAD THE UNIQUE POSITION OF BEING IN DOWNRANGE AUDIO CONTROL WHERE I COULD HEAR THE RADIO COMMENTARY, THE TV COMMENTARY, AND ALL OF THE PRODUCERS, SIMULTANEOUSLY. DON'S COMMENTARY WAS FAR AHEAD OF ANY OTHER COVERAGE. EVEN THOUGH A TV PICTURE WAS ON THE SCREENS SHOWING THE CAPSULE COMING IN WITH A PROBLEM, NO COMMENTS WERE HEARD ON TV UNTIL A MINUTE AFTER DON BLAIR HAD TOLD US WHAT WAS REALLY HAPPENING. AS A MATTER OF FACT, DON'S COMMENTARY THROUGHOUT, WAS ONE TO TWO MINUTES AHEAD OF TV. THE OVERALL PRODUCER IN HOUSTON CAUGHT ON QUICKLY AND BEGAN TO TAKE HIS CUES FROM BLAIR, IN ORDER TO ADVISE THE NETWORKS ON THE EXECUTIVE COORDINATING CIRCUIT. OBVIOUSLY, BLAIR HAD DONE MUCH HOMEWORK AND HIS LIASSON WITH THE NAVY AND NASA WAS CLOSE AND ACCURATE. ALL IN ALL, A VERY COMMENDABLE PERFORMANCE.

NASA, at one time, had planned on as many as five more Lunar landings after Apollo 15; but by the time 16 and 17 went up, explored new regions of the Moon each time and brought back more and more Moon rocks, the budget pressures were building up and people in high places were questioning the need for anything beyond Apollo 17.

That was where our love affair with the Moon came to an end on December 19, 1972, and the left-over Saturn V hardware became the building blocks of several spectacular exhibits. You will find one of them outdoors at the Johnson Space Center in Houston and another in the space museum at the Cape. It is monstrous anyway, but seeing it inside a building with each of its sections pulled apart from each other to allow us tourists to get up close and personal is absolutely awesome. The thought of getting that monster off the ground and the power it took is hard to grasp. At the very least, it inspires a whole new level of admiration for the scientists and engineers who were able to envision it, design it and build it.

By mid-June of 1973, however, NASA made good use of that other Saturn V when it began the Skylab program. The big rocket was needed to get the Skylab habitat into Earth orbit for the succeeding three manned missions to follow. Since those missions were primarily transports to the orbiting Skylab and not trips to the Moon, the full power of Saturn V was no longer needed.

A CONVERSATION WITH THE CAPTAIN

He was Captain Carl Seiberlich in July of 1969 and the proud skipper of the USS Hornet (CVS-12), which had been tabbed as the prime recovery ship for the Apollo 11 Moon mission only months before.

Now he is Admiral (retired) Seiberlich and living on a farm in Virginia, but definitely not sitting on the porch watching crops grow. His phone-answering message lets callers know that he remains an active businessman; and on the day he returned my call, he did touch on some of his present activities briefly, including his role in a new report on port security against terrorism. Obviously, the Admiral knows a few things about ports and port protection.

But our conversation centered on the Apollo 11 experience and I wanted to know how he found out that his ship, the Hornet, would have the honor of picking up the world's first Moon walkers. Three years earlier, Captain Gordon Hartley, then skipper of the USS Wasp (CVS-18) told me that his ship got the word from CNO, the Chief of Naval Operations. It was not quite the same for Captain Seiberlich.

He said he could almost remember, word for word, the comment of the Commander of the Naval Air Force, Pacific Fleet, who told him that with the added equipment and demands of a space mission that had to guard against possible lunar contamination (quarantine trailers and related protective devices) NASA and the Navy needed a ship with more than one screw, for starters, and additional space. I presumed he meant a larger hangar bay. As for single screw, that immediately ruled out the LPH's which had been active in the program only recently. Those

ships, such as Guam, Guadalcanal and Okinawa had been designed as twin screws, but, under budgetary pressures, were scaled back to one; and their critics took to referring to them as "McNamara's mistakes" with reference to the Secretary of Defense who presided over those

The NASA team aboard the Hornet after the Apollo 11 Command Module had been completely deactivated and all its contents removed and packed away for shipment to Houston. Not shown is NASA co-team leader Don "Doc" Stullken who spearheaded the flotation collar development during the early Mercury missions. His on-board partner Dr. John Stonesifer is at the far right in the white shirt while NASA Public Information Officer Ben James is the man in the white shirt on the left. Ben was the chief interface between NASA and those of us in the news media. (Courtesy of NASA)

changes. In addition, an LPH was hundreds of feet shorter than an Essex or Ticonderoga class carrier and could not accommodate fixed-wing aircraft.

As it turned out, Hornet was not involved in any extensive naval exercises when Apollo 11 rolled around, and Captain Seiberlich acknowledged that availability had a lot to do with any ship being selected for a recovery mission in those days.

I asked the Captain who picked the units that sailed with him on Apollo 11 and 12. "The helicopter squadron (HS-4) was a part of his ship," he said, "so no selection process was involved." He had nothing to do, however, with assignment of the UDT Teams which were UDT-11 on Apollo 11 and UDT-13 for Apollo 12.

How many recovery rehearsals would take place during a recovery cruise and who made that decision? I already had a pretty good idea that it was the Captain's call and he readily confirmed that. His goal, he said, was to ensure that when the real recovery day arrived, he could say that three people were trained and ready to do any one job in case there were injuries or illnesses. Obviously, NASA officials on board were more than willing to work with the Captain in order to get the recovery process perfected.

Therefore, we would be witness to partial rehearsals in which certain phases of recovery were concentrated upon. Some would be more for the benefit of the frogmen and in varying sea conditions, while others would be carried out at different times knowing that a spacecraft might return to Earth earlier or later than had been expected if problems arose.

As the Captain spoke, I could again hear the words of Chris Bent, Officer in Charge on Gemini 6 and 7. "The key," said Chris, "was practice and – we did, we did and did."

Did he have anything to do with saving the Hornet - one of only five legendary carriers of World War II, pre- or post-war, still around? There was obvious pride in his voice when he told me that he had a lot to do with it; and, he added, he had served as Executive Officer on the Intrepid before he assumed command of the Hornet and, while there, had participated in the recovery planning procedures for the Gemini and Apollo missions which were then still ahead. Hornet is now moored at Alameda Point in the San Francisco Bay Area and has an excellent Apollo recovery exhibit including a real MQF. Intrepid, as many of you know, is one of the other "saved" carriers and can be seen docked on the Hudson river in New York City. Yorktown is the the third and rides at anchor at Charleston, South Carolina. Two other great ships, which did not become involved in the space program, also survive as floating museums and both would be worth your while to visit if ever in their neighborhoods. Midway can be found and boarded in San Diego while the Lexington is on the bay in Corpus Christi, Texas. It makes a nice balance for anyone interested in Naval history. Two on the east coast, two on the west coast and one practically halfway in between.

As we concluded our lengthy conversation, Carl Seiberlich said he had two primary concerns in approaching the Apollo 11 recovery — protecting all of us, including the President of the United States, from "Moon germs," and making sure none of those priceless Moon rocks got away. Hornet was over 10 miles upwind of the Command Module at splashdown; and while I was a bit disappointed because I could not describe the descent in my radio broadcast, Seiberlich still recalled how he was taken to task by my television colleagues back then. I was not tuned in to them so I did not hear any such comments at the time, but obviously the three of them did not know that Seiberlich deliberately put Hornet that far away from the predicted splashdown point to provide an additional margin of safety with regard to the "Moon germ" issue. He knew, he said, that the rescue helicopters and their UDT Teams would be on scene within seconds of splashdown and they were.

Another issue came to light before we ended our talk. The Captain originally wanted to bring the Apollo 11 crew aboard inside the Command Module; but because an earlier mission had ended with the CM being banged against the side of the carrier, NASA wanted the men out of the spacecraft before a ship's crane brought it in. The skipper lost that fight, but he brought Hornet alongside the spacecraft while it was still bobbing in the Pacific and gave President Nixon a chance to witness their exiting the Apollo with their BIG suits on. Mr. Nixon wanted to go down to the hangar bay and shake hands with Armstrong, Aldrin and Collins, not realizing the urgency of getting the men inside the MQF ASAP and with no physical contact with the President, Admiral, or anyone else. After the Captain explained the potential high risks involved, the President backed off on his request.

Following our phone conversation, I received a copy of Captain Seiberlich's letter to the families of Hornet's crewmembers, sent to me by Hornet Museum curator, Bob Fish. Here is most of that letter.

USS HORNET (CVS 12)
FPO San Francisco 96601

29 July 1969

Dear HORNET Family,

Here we are on our way home, and by the time you receive this letter we should be back in Long Beach. I imagine that most of you watched our activities on the television, however, that told only part of the story. The long hours of practice, particularly those recoveries at night in the rain, were quite difficult. During the practice sessions we trained enough personnel so that we were three deep in all positions. Fortunately, we did not need this insurance. As you could see, these many practice sessions paid off handsomely on the big day.

Each of you have reason to be proud of your husbands, fathers and sons. Each man in this crew performed his duties in an outstanding manner. The sum of these performances was a flawless recovery. I would like to pay particular tribute to our Engineering Department. They kept this ship operating at top efficiency, throughout the cruise, generally in temperatures as high as 140° in the fire and engine rooms. My hat is off to them. Every other department performed just as efficiently, but I mention the engineers because of the difficult conditions under which they worked, and also because they saw little of the recovery operation.

One of our customers, Astronaut Neil Armstrong, after being picked up said, "we are most pleased with the recovery, it was the smoothest and most professional operation we have ever witnessed or participated in." Admiral John S. McCAIN, Commander-in-Chief of all U. S. Military Forces in the Pacific sent the following message to HORNET:

"1. Yesterday I was privileged to join President Nixon aboard HORNET to witness firsthand the recovery of our Apollo 11 Astronauts. The completion of this historic mission, another significant milestone in the progressive heritage of our great nation and one in which we can all be justly proud.

2. The President personally expressed to me his appreciation for the exceptional performance of every Bluejacket, Soldier, Airman and Marine who participated in the flawless execution of the entire Apollo 11 mission.

3. I wish to add my personal commendation to that of the President. The selfless performance, dedication and professionalism of each individual participant in the recovery Task Force contributed significantly to the success of the epoch achievement. I commend all hands on a job exceptionally well done. Admiral John S. McCAIN, Jr., U. S. Navy."

Mr. J. B. Hammack, Chief of Landing Recovery Division, National Aeronautics and Space Administration sent the following message to HORNET, "This is to officially express my heartfelt congratulations for the great recovery job done by the USS HORNET. I was fortunate to observe the recovery operation personally and want you all to know how deeply impressed I was at the great performance of your fine ship. I have never been on a ship that had greater morale and dedication to purpose than USS HORNET."

Many other messages were received from other high ranking officers such as Vice Admiral SHINN, Commander, Naval Air Force Pacific Fleet and Rear Admiral TRUM, Commander ASW Group FIVE, however, I wanted to quote the few above which are typical of the high opinion in which we are held. This praise was earned through long hours of hard work by every man on this ship.

You also made your contribution to this operation by holding up your end at home. Often this is more difficult than actually doing the job. To our HORNET family, a "WELL DONE".

Sincerely,

C. J. Seiberlich

C. J. SEIBERLICH

The letter went on to detail the ship's upcoming activities back in and around Long Beach, California.

Having witnessed five recoveries and the tireless preparations that went into all of them, I can attest that this letter was not just posturing on the Captain's part but a sincere expression of appreciation for a job very well done. The fact is, a letter of this nature could have been dispatched after each and every splashdown, and probably was.

THEY'VE GOT A GIRL IN THEIR ROOM!!!

It happened on my very first downrange splashdown assignment. I split the recovery day broadcasting with the veteran Bill Ryan of NBC-TV, New York. In the days leading up to the splashdown, Bill would file his daily radio reports from the little Mutual van behind the Wasp's island plus a similar report for CBS. I filed for Mutual (my network) and ABC. This was standard procedure.

What was not standard was the little caper the ABC guys pulled on a couple of naïve Philippine cabin boys. It seems that in that era before the U.S. cut the Philippines loose from U.S. control, (they were a U.S. territory) the military, or at least the Navy, had a policy that any Philippino lad signing up for a four-year hitch in this man's Navy would automatically become a U.S. citizen at the end of that enlistment. Quite a prize in those days; and, besides, the Navy pay, even then, far exceeded what most of them could have been making back home. One example was pointed out to us. A robust Chief Petty Officer (aren't they all robust?) who had been sending part of his pay back to his village for years was not merely the wealthiest member of his community, but, according to some of his shipmates, he practically owned the village. Possible.

Anyway, these ABC folks, given to television excellence, always mixed with a fair degree of fun and games, decided that their loyal cabin stewards were ripe for some hi-jinks. While it may look like out and out discrimination in that so many of those Philippino lads were put into positions you might expect in a hotel, many of them were still struggling with our language and it probably was the smartest, if not the easiest way, to assimilate them into the system.

Nonetheless, the TV guys went ahead and hatched their little plot. Having watched these reliable young men bring them fresh towels and

such for several days, the ABC bunch headed down to the sea store aboard the Wasp one day and got themselves some ladies nylons, a pack or two of women's panties (the kind with the day of the week on each pair) and a red marking pencil.

The stage was set for deception. On the selected day, the ABC lads who were bunked about 6 to a room, draped those nylons over their locker doors, strategically placed any number of women's panties on their bunks, and put a few squished out cigarettes in the ashtrays with little specks of red around the filter tips.... ala lipstick. Got the picture? Down the hall come the Philippino lads, their arms full of towels, washcloths and the like. As they approached, one of the ABC guys lets out a yell, which was a cue to slam a few locker doors as though they really had something to hide. Enter our Philippino friends, who immediately began looking around the room while the ABC guys did a fairly good job of trying to look cool and innocent. While the stewards said very little, their faces said a lot – a whole lot. As later described to Bill Ryan, myself and others who could not be witnesses, the young men were nothing less than frantic, and one could only guess what they must have been thinking. They deposited their towels and made a hasty retreat, babbling in their native language as they sped back down from whence they came. It seems apparent that they left that room that day fully convinced that the American TV people were nuts enough to actually smuggle a girl into their room and that she had stepped in to one of those tall steel lockers moments before they entered. They probably told the tale for a long, long time, and that was just fine with the ABC crew. As I said, they do excellent work and are never less than professional when cameras are rolling and things are happening. When the little red light is off, they are a fun bunch.

THE RED CARPET TREATMENT (GT-11)

During the five manned recovery missions I covered as a "pool" correspondent, there were a handful of memorable pranks – always done in the spirit of good fun and never harmful to anyone. But when two high ranking NASA officials and an Admiral joined forces to have some fun with a Marine Major, at his expense that is, it was the first time I was witness to anything with that sort of high-level sanction, but the motivation was basically the same as it had been on all previous or succeeding occasions. It was a means of relieving the sometimes boredom of a long sea voyage and the repeated simulations for the real manned spacecraft recovery.

The setting was on board the USS Guam and the mission was Gemini 11 in September of 1966. There was a Marine platoon to perform various roles – color guard, security, flag bearers, etc. Their commander was a Marine Major who showed us very little sense of humor but total dedication to his job and to the young men he directed.

A daily ritual was the morning physical fitness routine the Marines went through. Many, many laps of Guam's flight deck, many, many push-ups and calisthenics, etc. At the end of each evolution the Major, standing before them, would ask the Marines what they wanted next. The answer was always the same. "We want more Sir!" And off they would go again…running around the flight deck.

The top two NASA men on board were daily witnesses to this tribute to fitness. I don't believe NASA team leader, Dr. John Stonesifer, or Public Affairs Officer, Ben James, had any problem with physical fitness. They could watch it all day like many of us do. They were somewhat bemused by the gung-ho young men among us, however, and their hard-nosed Major. They decided to play a trick on him, and by the time John told me what they were up to, they already had the cooperation of Rear Admiral Roy Anderson, commander of the Westlant (Western Atlantic) Recovery Force. The Admiral apparently had a much broader sense of humor than our Marine Major friend.

Gemini 11 splashdown day, our Marine Major is front row center, and the red carpet was rolled out without further incident.

The Major kept a tight rein on a roll of red carpet that was unrolled during each simulated spacecraft recovery. It would stretch from the steps of the recovery helicopter down to the Guam's island amidships where microphones were set up for the ship's Captain and NASA officials to offer their welcomes and for the astronauts to thank the Navy for another sharp recovery operation.

Well, one day there was a rehearsal but no red carpet. Of course, the Major was royally ticked off and he looked around until he locked onto his NASA fans looking down from the island's upper levels and pointedly warned, "I'll get you." The Major bent every effort to keep that precious red carpet under lock and key, but when you have an Admiral working against you, you're licked. Our conniving NASA duo had access to it.

Somebody got a hold of a nice sheet of poster board, and the night before one of Guam's final simulations, there was a neat bit of business going on in the Admiral's cabin. Three grown men on their knees, cutting, pasting, coloring what became a very large human hand with the middle finger extended – the sort of gesture every living creature on this Earth understands.

You can probably tell the rest of this tale yourself. That big hand found itself into the roll of red carpet where it remained until the recovery helicopter touched down on Guam's forward deck and two "pretend" astronauts prepared to step down onto the carpet.

That was the cue for the big rollout. Kick the carpet and off it went — but midway through the unrolling out came the hand with the middle finger raised for all to see. Even those young Marines had to laugh, although it might have cost them a few hundred push-ups. The Major did not laugh. He slammed his hat down on the deck and immediately looked up at the island's 06 level where all the culprits were standing and laughing heartily. I heard the uproar clearly down at the stern where I was interviewing a GTE technician at the giant antenna that company had installed for the real recovery.

I understand the Major had some more choice words for his NASA fans. To this day, John Stonesifer, now comfortably in retirement like so many of us, insists it was all in good fun, and I doubt that the Major did anything more about the trick played on him than fuss and fume for a while. What else could he do? An Admiral was a major player in the Red Carpet Caper.

The real recovery and deck ceremony took place a few days later without incident…and without "the hand."

THE EQUATOR BECKONS (APOLLO-11)

There comes a time in the lives of most sailors when the ship they call home crosses the equator in one ocean or another, and there is a price to be paid if that crossing happens to be your first. It's a tradition, barbarism to some, that has been going on for as long as…. well, as long as naval vessels have been crossing the equator, and the odds are that in its earliest manifestations it was a helluva lot worse than it is today. I might argue that point.

When we left Pearl Harbor several days before Apollo 11 began racing to the Moon, the last thought in my mind was an equator-crossing ceremony. I would be a long time getting over the fact that I was actually where World War II began and bemused by the presence, right across from Hornet's pier space, of a Japanese submarine. One of the first things my roommate, Ron Nessen, said to me as we stood at a hangar bay door and watched the tugs nudge us out into the channel at Pier Bravo was, "As long as he (the Japanese sub) is in here, we've got nothing to worry about." Just a hint of the droll humor we were to hear a lot of in the next two weeks.

Then we were on our way, but not before solemnly passing by the Arizona memorial where over a thousand sailors remain and where tiny drops of oil continue to rise to the surface even today. It is a sight which never fails to inspire reverence.

Our course was south-southwest from Hawaii and, naturally, had to take us across that equator. We were forewarned at least a day before that those of us who would be crossing for the first time were mere low-down slimy pollywogs who would need some straightening out on the mysteries of the deep, and our low position on the world stage was spelled out clearly in the summonses all of us "equatorial virgins" would find in our cabins before the big day.

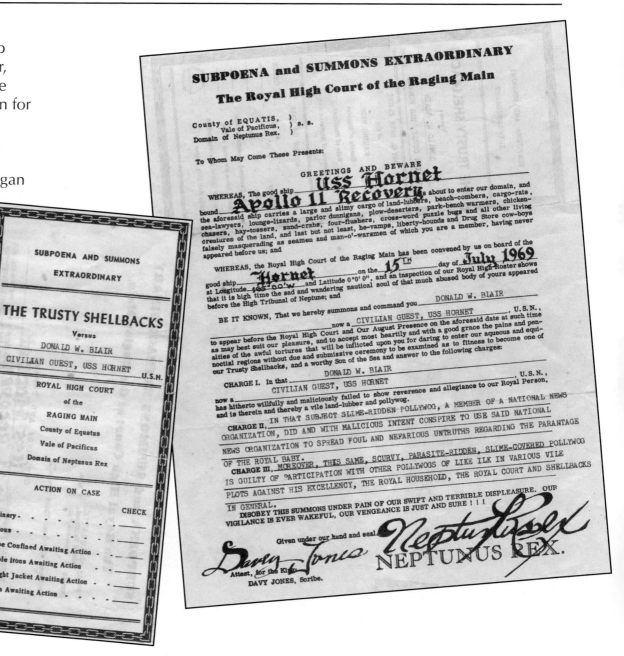

Subsequently, we had to appear before the High Tribunal of Neptune. I guess we also had to kiss the Royal Baby's belly (hey its been 34 years at this writing…give me a break) and if we had any wretched reasons for not wanting to submit to the crossing ritual that was the time and place to fink out. There were hundreds of us but only one decided to pass up this great chance at human humiliation, and he was a reporter from the New York Times. While he shall remain nameless, (he went on to ABC Television) he did comment that he thought the whole thing was "barbaric." They let him off the hook.

The rest of us decided to take our medicine, including my late friend and mentor Dallas Townsend of CBS. Dallas had made the crossing on a Navy ship during World War II and did not have to go through all that nonsense again, but, not having his little wallet-sized card proving his legitimacy, he was too much of a man to beg off. While always much-admired in our business, he grew just a bit bigger in my eyes after that.

The ritual is a pain – in the knees and some places you didn't even know you had places. First of all, it is done on your knees. No standing anytime. My recollection is that we traveled the decks and ladders

One of the first things you learn as this ritual unfolds is "keep your head down." It is forbidden to look up because you are surrounded by whip-wielding sadists who have gone through what you're going through and can't wait to give something back. (Don Blair photo)

(stairways?) of the USS Hornet for something like four to five hours. Try climbing a ladder on your knees sometime and you come close to feeling what we felt. Up a nearly vertical stairway, or whatever the Navy calls those things, and out onto the flight deck, looking for the entire world, like a herd of cattle but not being treated as kindly.

Look up and you will quickly be visited by a young man carrying a whip made out of (usually) parachute straps, and he will then whip your poor butt until you decide that you will never look up at anything again. I looked up because one of my dear NASA friends called my name and said he wanted to get a picture of me. I couldn't resist but I've been wishing I had ever since. The blood bruises on my rear end did not go away for weeks.

Mind you, our "herd" of slime-ridden pollywogs consisted of news media folks like myself, some of the NASA guys, and plenty of first-time-over-the-equator sailors, helicopter pilots and crewmen, and UDT (frogmen) members. The frogmen – Navy SEALS these days – are probably some of the toughest fighting men on Earth and they don't shrink from anybody. I guess it came as no surprise to any of us when a group of them did raise their heads to serenade those Marines among the whip-wielding Shellbacks (those who have lived through a crossing). Their song? "I'd rather be a UDT than a candy-assed Marine!" I'll never forget it and I wouldn't be surprised if their bruises were still showing.

Now keep in mind that we were not merely moving around the ship on our knees. They had lots of little treats in store for us, and I won't even try to recall the order in which these goodies were handed out. At one time or another we would slither through a long canvas sleeve that was kept oozing with kitchen slop that they had saved up just for us. Loads of fun. We were all directed, somewhere in the proceedings, to get into a wooden coffin whereupon the attendant Shellback would slam the slid down and beat on it for perhaps 15 or 20 seconds while asking how you were doing.

Any perceived violation of the rules and regulations of initiation would also bring the Hornet's fire hoses, which are normally used to squelch aircraft fires or anything else that catches fire aboard ship. Very high pressure, needle-like spray which does wonders for your skin tone and like those whips, leaves a lasting impression.

It might have been when this Do-Do bird lifted his weary head to have his picture taken that the same Shellback who tenderized my tush

also took the time to pull out a hefty looking pair of scissors and relieved me of a precious lock of hair right up in front. Took about three months to grow it back.

I would also note that sometime during our long stay on the flight deck, there were times when we thought they might decide to brand us like cattle in some fashion. After all, we certainly looked like we were on our way to a market somewhere. But I'll never forget a guy right next to me who frequently went to his Rosary beads and muttered a prayer now and then to help him get through it all. Fortunately, nobody bearing a whip felt that his gesture warranted further beating.

Oh, the things that little boys do to little boys. Many critics feel just that way and only once during the whole ordeal did Captain Carl Seiberlich see fit to dispense anything resembling mercy. From the bridge we heard his verbal order to cool it with the fire hoses, but that was more to do with the always-critical fresh-water situation on a ship than any sincere act of mercy on the skipper's part. He was probably visualizing the unusual amount of water we would all need in the showers later as we tried to shampoo out the gobs of aviation grease that was so generously rubbed into our hair throughout the fun and festivities.

Hey, we lived through it. No fatalities. Just some sore asses and a wonderful certificate suitable for framing and bragging back home. Take a look. They got my middle initial (W) wrong but I decided never to mess with it. Leave it as it is. July 15, 1969 – the day I became a real guy and went from Pollywog to Shellback. It was a beautiful thing. You had to be there.

SO WHO BUYS THE DRINKS, ANYWAY?

Here's another old Navy tradition and one based strictly on your experience with the business of crossing the equator. If you are one of those poor unfortunate souls who has not been initiated into the secrets of the briny deep and Neptune's world, then you are a slimy Pollywog and there must not be anything worse than that in the eyes of a Shellback. A Shellback is, obviously, the opposite. One who has made the crossing and paid the price.

So now you can imagine the scenario where two sailors meet and one of them asks, "Are you a Shellback?" There is only ONE acceptable answer, and, as a matter of honor code, it would be the ultimate in bad judgment to try to lie your way around it. If you haven't crossed the equator you buy the drinks. Get used to it. But if you are of that fraternity (and I assume, have a wallet card proving it or can simply say so and look sincere while doing it) you respond by filling in the words indicated on the Shellback patch by their first letters. "You Bet Your Sweet Ass I Am." Case closed.

Incidentally, when the crews of several of the Apollo 11 recovery helicopters invited me to join them at a popular Honolulu steak house prior to our sailing, they introduced me to the squadron drink, the Black Frog, and I still enjoy one occasionally 34 years later. Two parts scotch and one part Kahlua…. on the rocks. Cheers!

He was too young to be called a curmudgeon but he did have salt & pepper hair so perhaps he was getting close. My recollection of Sid Andrews was as the omnipresent cable puller, trailing the ABC cameras wherever they went on the USS Okinawa during our Apollo 15 cruise. Sid probably could have made a fair living as a Catskill's comic – maybe an opening act for some headliner of the time. Always ready with the one-liner but never to the detriment of tasks at hand.

Early on, we of the media were introduced to his "shtick" when we would assemble in the pilot's Ready Room for a daily briefing on the latest recovery rehearsal, the weather outlook, and anything else we might take note of and report on in our daily radio filings.

From down the hall we would hear a sharply spoken 'TEN-SHUN!' and the whole room, civilians included, would stand up and snap to attention out of courtesy to the skipper. In the next moment, a laughing Sid Andrews would enter, shout "At ease," and the enjoy the moment, or two before Captain Andy Huff – totally in tune with the joke – would walk in and we would all sit down, still laughing.

I clearly recall how much Sid's antics loosened everybody up and contributed mightily to the feel good atmosphere that was so much a part of our time aboard Okinawa.

Sid hatched a plot against the well-liked ship's supply officer, Lt. John Bian, when, at a subsequent briefing, he read a list of charges against Bian, including overfeeding all of us, as Bian stood before us head down but with a barely suppressed grin on his face. A few days later the Chaplain (talk about your conspiracies) had the Lieutenant arrested and led right out of the Officer's Mess. He had his dessert in the brig.

At another time during the cruise, the ship put on a talent show which Sid had a major role in assembling and directing. On a bright, sunny day in the Pacific, north of Honolulu, Okinawa's port side elevator rose slowly from the hangar bay to the flight deck where hundreds of us waited as an elevator full of singers and musicians appeared before us. It was an impressive sight and some considerable talent on display. Toward the end of the show Captain Andy Huff asked Sid to please come forward to be thanked for all his help and to laud his many contributions to the mission. Sid played his role to perfection, staring straight ahead as Captain Huff accepted a foil-covered plate handed to him by my "pool co-correspondent" David Horwitz of NBC Los Angeles. The Captain asked Sid if he was ready to receive his "award" and Sid replied, "Yes sir." Whereupon the Captain unloaded a whipped cream pie dead center on Sid's face. You could have heard the roar back in Pearl Harbor.

Everybody on that ship knew Sid, liked Sid, and the feeling was mutual.

But the Gremlin wasn't done yet. Although he was not the technician who actually did the deed, there came a moment at the excellent splashdown dinner (best food ship I ever sailed on) when, with Sid's certain connivance, Captain Huff and NASA Team Leader Dr. Don Stullken stepped up to an impressive looking layer cake upon which stood a model Gemini-Atlas rocket. The lighting in that part of the dining room had been purposely dimmed for what was about to happen. There was a fine wire attached to the nose of that rocket, and I do not believe the Captain and Doc Stullken were aware of it. The moment the Captain reached in to cut the cake, that rocket took off and hit the ceiling. I took pictures of that incident but, not having packed flash equipment for the mission, they did not turn out very clearly. We tried. You should have been there.

Diversions at sea. There were many on the missions I was privileged to take part in. Riding a highline from ship to ship, convincing a pair of Philippino sailors that a room full of ABC techies had a girl in their room, giving a by-the-book Marine Major the famed one-finger salute, crossing the equator and paying the price for it, arresting a Supply Officer and giving him some brig time, a pie-in-the face ceremony and a cake that almost went into orbit. I might have left out a few and I still wonder what sort of hi-jinks might have taken place on all those carriers I did NOT sail on. No, Bob Hope never helicoptered in to entertain us, but he might have given it some thought during the Apollo 11 mission. Now that would have been a photo opportunity. Instead, we of the media, with a ready-to-have fun Navy right up to and including a Captain and an Admiral, had to conjure up and carry out our own merriment. I think we did a helluva job.

Frogmen of UDT-11 dropping out of a helicopter from squadron HC-1 during a recovery rehearsal for Apollo 15 in August of 1971. (Don Blair photo)

An HC-1 squadron helo hovers above the Okinawa's boilerplate capsule, now with collar deployed and UDT-11 raft attached. Okinawa painted their boilerplate capsule white and added the flag. Nice touch. (Don Blair photo)

At right: For the first time since men landed on the Moon, this mission did not require the crew to be ushered into a quarantine trailer aboard ship. The Apollo 15 crew of Dave Scott, Jim Irwin and Al Worden, therefore, could revert to the older method of gathering on the flight deck before a microphone to thank the recovery ship for one more great job. (Don Blair photo)

USS LAKE CHAMPLAIN (CVS-39)

The first primary recovery ship for a manned U.S. space flight – the brief, sub-orbital flight of astronaut Alan B. Shepard Jr. on May 5, 1961, and the second flight in the two-man Gemini series – astronauts L. Gordon Cooper and Charles "Pete" Conrad Jr. on August 29, 1965.

The Lake Champlain was one of those modifications of the Essex class carriers which were lengthened and became Ticonderoga class. The most visual difference was reshaping of the bow to allow for two 44mm quadruple gun mounts – a vital need for anti-aircraft capability in the South Pacific during WWII. Another characteristic shared by all those that became Ticonderoga class was their length – 888 feet. In time, 13 ships would become Ticonderoga class by the end of 1946.

Curiously, while the Lake Champlain and her kin would be as much as 16 feet longer than the so-called "short hull" Essex class, her weight was almost identical at the time of launching – just over 27,000 tons. She had a beam of 147'6" and could attain a top speed of 33 knots. She was built to accommodate 3,448 officers and crew.

(Navy photo)

A BRIEF HISTORY OF THE USS LAKE CHAMPLAIN

Her keel was laid down at the Norfolk Navy Yard in Portsmouth, Virginia on March 15, 1943. She was launched on November 2, 1944, and commissioned into service on June 3, 1945.

The war in Europe had ended with the unconditional surrender of Nazi Germany on May 7, less than a month before Champlain's commissioning, and victory in the Pacific was near at hand although few Americans had any idea that a device called the Atomic Bomb would hasten that day considerably.

Consequently, the Champlain never launched a warplane in anger in that conflict, nor did it ever have to worry about Japan's kamikaze pilots, as did so many other U.S. warships in the Pacific.

Instead, her only involvement with World War II was to put into South Hampton, England, on October 19 where she embarked many of our veterans and brought them back to New York during Operation Magic Carpet.

Barely six weeks later she set a trans-Atlantic speed record of 4 days, 8 hours and 51 minutes, dashing from Cape Spartel in Africa and heading for Hampton Roads, Virginia. That record held up until broken by the luxury liner SS United States in the summer of 1952.

The Korean War brought the Lake Champlain and many other ships of the line out of mothballs at Norfolk for reactivation and modernizing in the fall of 1952. It was at this time that she was redesignated an attack carrier and given the number CVA-39. She moored at Yokosuka (pronounced YO-KOOSE-KAH) Japan, on June 9, 1953.

Her participation in the Korean conflict was exciting, action-filled and brief as her air groups began attacking enemy airfields and ground forces on June 14, escorting B-29's to their targets and carrying out numerous other air strikes until the truce was signed on the 27th of July. She was back in the U.S. before the end of the year, arriving in Mayport, Florida, on December 4.

There was action with NATO forces in late April of 1957 when the Lake Champlain joined elements of the fleet in a fast run to the Middle East where she helped to ease Jordan's stand against the threat of Communism.

By August 1 she had been converted to an anti-submarine carrier and reclassified CVS-39. A lot of the needed conversion work was accomplished at shipyards in Bayonne, New Jersey. However, while many carriers of her era (many of them Essex class) were modified with an angled deck, plans for a similar upgrade for Champlain were canceled.

Life got a bit quieter for the big carrier until she was notified by CNO, the Chief of Naval Operations, of her designation as the recovery ship for America's first manned space effort. She sailed for the recovery area on May 1, 1961; and, after numerous spacecraft recovery rehearsals (SIMEX's), recovered America's first man in space in near-flawless fashion.

Fast-forward to the 24th of October 1962, and the USS Lake Champlain played an important role in what was probably the most tension-filled encounter of the Cold War era – the Cuban missile crisis.

She was one of many U.S. warships whose task was to block any vessels, Soviet or otherwise, thought to be bringing offensive missiles to Castro's island nation. The blockade, of course, worked and the high tension evaporated.

Back to anti-submarine work in waters off New England, but not too busy to pitch in and help out victims of hurricane "Jane" when it smashed into the island of Haiti in September of 1963.

What many of her officers and crew may have regarded as icing on a celebrated and honorable cake came when Champlain was tapped for one more astronaut recovery mission – the flight of Gemini 5 in August of 1965. Once again the Lake Champlain was up to the assignment and astronauts L. Gordon Cooper Junior and Charles "Pete" Conrad found themselves in good hands following what had been America's longest orbital flight to that date – 120 orbits of the Earth consuming just under 191 hours. This time, however, Champlain was not first on the scene.

The ship was not as close to the descending Gemini capsule as many others would be in the years to come but (and this happened several times during the so-called "splashdown" years) a back-up ship – the destroyer USS Dupont (DD-941) raced in to recover the crew and then transferred them via helicopter to the Champlain's flight deck. This set the stage for what some might consider a controversy. More on this in the chapter on the Gemini 5 mission. Despite the question marks, Conrad and Cooper stepped onto the carrier's deck 89 minutes after splashdown.

Not a bad way to end a fine Naval career. Soon after Gemini 5 she sailed for Philadelphia where she commenced her deactivation. She was decommissioned on May 2, 1966, and spent about six years in mothballs before being sold for scrap. As did many of her sister ships, she succumbed to the welder's torch at shipyards in Kearny, New Jersey, in 1972.

Not exactly a giant step for mankind. That would come over eight years later. But Mercury-Redstone 3 was a giant step for this country, still smarting from the Soviet Union's shocking launch of the Sputnik 1 satellite almost four years earlier, and the first man to orbit the Earth (Yuri Gagarin) mere weeks before Shepard would make his brief, sub-orbital flight.

Alan Shepard was one of the original seven astronauts, the seemingly too-good-to-be-true ace pilots hand-picked from the Marines, Navy and Air Force. Shepard was a Navy Commander at the time.

Television, radio and the print media combined to bring our tensions and anxieties to a peak that day in early May. For months, as NASA flexed its new found wings and mandate, we were treated to insights into the lives and training regimens of these very special men. They were cover boys for Life, Look and many more of our most popular magazines. All-American boys. Only much later would the world know, mainly through books written by themselves and others, that a few of them liked to party and raise hell now and then.

But on that May 5 morning all eyes were on a long-since discarded launch pad at Cape Canaveral and television coverage ruled the moment. The anchormen and space specialists at all three of the major networks had schooled us well in the dangers and uncertainties facing our newest heroes. We were rarely allowed to forget that Alan Shepard and those to follow were literally sitting on top of a very volatile explosive device designed to send deadly missiles into space – not humans.

In the years since Sputnik 1 there had been many unmanned attempts to send our hardware into space, and many of them had exploded on their launch pads or had to be destroyed shortly after liftoff. Critics and comics were having a field day. Shepard's rocket had been built by "the low bidder." Something to think about as you sit on top of it in your cramped little Mercury spacecraft. Others said the U.S. space program was like Civil Service. You couldn't get it to work and you couldn't fire it.

But here it was May 5, 1961. Time to put up or shut up. We couldn't see the beaded brows at Mission Control nor the sweaty palms, but it was clear those conditions existed. Verified by dozens of books to come in the ensuing decades.

The countdown. Destined to become a regular nail-biter for all of us in the coming years. And then suddenly, noisily, at 9:34 a.m., Shepard's long, slim limo was lifting off and a nation held its collective breath. Up, up and gone and soon out of sight with just Mission Control to maintain communication with the astronaut and those TV folks to assure us that things were going nicely.

It was short, and looking back, very sweet. Just a tad over 15 minutes and we could stop holding our breaths. Splashdown. A first for the first American in space and a veteran aircraft carrier, the USS Lake Champlain, which became the first primary recovery ship for a manned mission. They continued to be aircraft carriers for very obvious and practical reasons but a few destroyers would get into the act before splashdowns gave way to the Space Shuttle in 1981.

When it was over, some 300 nautical miles downrange in the Atlantic, Lake Champlain had been one of 10 ships in the recovery fleet together with 7 fixed-wing aircraft and 7 helicopters. Shepard was in the good hands of one of those helicopters provided by a proud Marine helicopter squadron, HMR (L)-262, part of MAG (Marine Aircraft Group) 26 out of New River, North Carolina.

Freedom 7 had attained a maximum velocity of 5,134 miles an hour while soaring to an altitude of just over 116 miles and creating an arc that ended in the Atlantic just over 300 miles downrange. Total flight time – 15 minutes, 28 seconds.

Alan Shepard had shown his NASA compatriots that a human being could stand the rigors of extraordinary G-forces, weightlessness (although for only about 5 minutes) and re-entry, and that he could perform just about every action and maneuver necessary to ensure a safe, successful ride. Man could and did function in space. It was enough to spur President John F. Kennedy to proclaim just three weeks later, the goal of "landing a man on the Moon and returning him safely to the Earth before this decade is out." That proclamation left many observers in shock, including many in the space program itself, but, as we all know, NASA was more than up to the challenge.

Shepard's Mercury spacecraft, Freedom 7, can be seen in the lobby of the Medical Center at the United States Naval Academy in Annapolis, Maryland. At the time of his first flight, Alan Shepard was a Navy Commander.

Before we continue on into the Grissom flight and others, this is the right moment to look back and refresh ourselves on this fascinating new world we were getting into.

Long before titanium became the favorite metal of the world's golf club makers, it emerged as some sort of "super metal" in the NASA scheme of things.

The Mercury capsule frame itself was made of titanium but covered with steel and beryllium "shingles" as they came to be known. At the outset, what we came to call the heat shield was a heat sink made of beryllium. Later flights introduced the heat shield and the word ablative. Now we were learning about a material designed to absorb thousands of degrees of heat upon re-entry into the Earth's atmosphere and subsequently burn away, at least partially, while keeping the spacecraft itself safe from that intense heat.

The Mercury capsules were made by the McDonnell-Douglas Aircraft Corporation, and that St. Louis company would go on to build the two-man Gemini spacecraft.

Small? The one-man Mercury was just over 9 feet tall and 6½ feet wide. At launch, with a lot of gear stowed, it weighed in at 3,650 pounds but it is only a ton and a half when you see one in a museum these days.

NASA started out with the pure oxygen atmosphere in its spacecraft and that would be good enough until the Apollo launch pad fire on January 27, 1967, which claimed the lives of astronauts Grissom, White and Chaffee.

Note: These mission summaries are not intended to "dot every I nor cross every T" as has been done expertly in the shelves full of books authored by NASA astronauts and operating officials. Those fine volumes describe the ups and downs of all these missions in exquisite detail and I wouldn't hesitate to recommend all of those which I have read and that is…most of them. For these pages, I am seeking to highlight advances and breakthroughs while focusing equal attention on the ships and men which played such important roles when those missions "splashed down."

USS RANDOLPH (CVS-15)

The USS Randolph became the second aircraft carrier to be picked as prime recovery ship for a manned space mission – the sub-orbital flight of Gus Grissom on July 21, 1961; and eight months later – the same assignment for the first American to orbit the Earth, John Glenn on February 20, 1962, although Glenn overshot the landing spot and the actual recovery was carried out by the destroyer NOA.

The USS Randolph was Ticonderoga class, which meant a length of exactly 888 feet as noted earlier (see – USS Lake Champlain). Her other "stats" were basically the same.

A BRIEF HISTORY OF THE USS RANDOLPH (CVS-15)

Randolph's keel was laid down on May 10, 1943, at the Newport News Shipbuilding and Drydock Company in that Virginia port city. Launch came on June 28, 1944, and commissioning on October 9 of that year.

It would be early 1945 before Randolph's hull would even touch Pacific waters but she wasted no time in dispatching her aircraft against Tokyo airfields and a Japanese engine plant by mid February as a member of Task Force 58. The war in the Pacific, while still furious, was winding down as the attacks on Japan itself vividly illustrated and even the island groups then under attack were in Japan's shadow, such as Iwo Jima.

On March 11, while anchored at Ulithi Atoll in the Carolines (some 1,300 miles south of Tokyo), Randolph was struck just below the flight deck by a Japanese twin-engine bomber, killing 25 men and wounding 106. She was repaired right there and joined the Okinawa Task Force in early April.

From that time on, Randolph shuttled back and forth between several Task Forces but her basic mission remained the same – attacks on targets in Japan and the nearby islands. Her planes were still hitting airfields and industrial installations in Japan the day that nation surrendered – August 15. For the rest of 1945, after transiting the Panama Canal and arriving back in Norfolk, she was rigged for the so-called "Magic Carpet" service to bring back American servicemen from the Mediterranean.

(Navy photo)

Placed out of commission in February 1948, she was berthed at Philadelphia but was put back on line on July 1,1953, after reconditioning at the Naval Shipyard there. Her new classification was CVA-15 – attack carrier, and she headed for the Mediterranean and duty with the 6th Fleet in 1954.

It was back to the states and Norfolk, Virginia, in 1955 and a lengthy stay in the shipyard there for installation of the angled flight deck and the enhanced "hurricane bow." Three more 6th Fleet deployments followed during 1956 to 1959 following which she was converted to an anti-submarine support aircraft carrier and reclassified CVS-15.

Over the next decade Randolph took part in ASW exercises in the Atlantic, Caribbean, the Med and off northern Europe while also acting as the primary recovery carrier for early Mercury manned space flights, starting with the sub-orbital flight of astronaut Gus Grissom on July 1, 1961, and the "first American in orbit" flight of John Glenn on Feb 20, 1962.

That summer the Randolph joined the huge fleet sent to isolate Cuba during that year's tense missile crisis. Then, another overhaul and, during the ensuing five years, two more Mediterranean and northern Europe deployments, winding up her naval career in the Caribbean.

Pentagon cost cutters were busy doing what they do in August of 1968 and Randolph was among 50 ships slated to be written off. She was placed out of commission, put in reserve and berthed at Philadelphia until June 1973, when she was stricken from the Naval Reserve Register and sold for scrapping.

For her service in World War II – three battle stars.

For ASW operations in the Caribbean – a fourth Battle Efficiency "E" in September of 1960.

First, the obvious. It was called Redstone 4 because it was the fourth mission in that series and it would be the last to use the Redstone rocket. All the Mercury flights had a seven after their spacecraft names in recognition of the astronaut's status as our original 7 space fliers.

The Grissom flight introduced us to the unpredictability of weather and its immediate and obvious effect on NASA's plans. While the Shepard flight experienced numerous "holds" in its countdown, that mission did fly on the day it was finally scheduled after a number of "glitches" were dealt with.

Grissom was up and down the elevator to his spacecraft several times due to weather delays but had to stay with Liberty Bell 7 during the many built-in holds while the launch team, in what would become familiar NASA fashion, checked and re-checked systems and equipment.

Those of us in the media who were piling up NASA press kits to facilitate writing and reporting on this great new American adventure were soon to become familiar with words like ablative (protective coating/heat shield) and hypergolic (spontaneous ignition when two unfriendly fuels get together). Redundancy, while a fairly familiar word, would become synonymous with NASA systems that would be backed up with one and usually two more of the same. In space when something fails, it's a little late to try and "take it back to the shop" so NASA tried to ensure that its astronauts could quickly and easily switch to the next computer or fuel cell in line or - whatever.

If you were nervous while watching a lift-off at home, imagine how the men in the spacecraft must have felt. On the day he became the second American in space, Gus Grissom sat in the capsule for nearly three and a half hours. It sounds like an eternity to be confined almost motionless in such a cramped space, but other crews would go on to surpass it.

Just as in the Shepard flight, the Grissom sub-orbital would provide further corroboration of man's ability to function in space, survive G-forces and re-entry, and operate all necessary equipment.

There were already some new wrinkles. The addition of a large viewing window and an explosively- actuated side hatch. Already the astronauts had lobbied for a better view than Alan Shepard had. As for that explosive hatch, it would prove to be a fateful change, at least on this mission. July 21, 1961. After several weather delays and what would become expected built-in "holds" for equipment checks, Grissom's Liberty Bell 7 lifted off for the second and last U.S. sub-orbital mission at 7:20 a.m. When Gus splashed down in the Atlantic, some 302 miles east of the Cape, it was almost a carbon copy of Alan Shepard's trip, coming within a few seconds and a few miles of that flight. Mission accomplished. Well – almost.

That's when a big-time "glitch" kicked in. Shortly after a Marine helicopter (from HMR-L-262) latched onto the capsule, the hatch blew open prematurely. Grissom jumped out as helo pilot Jim Lewis worked to move the now waterlogged capsule away from him. After several anxious minutes, another helo hoisted him to safety. With the capsule now exceeding the chopper's lifting capacity, it slipped away and sank in 15,000 feet of water.

Inevitably, questions were raised as to what, if anything, Grissom's performance might have had to do with hatch malfunction. Nothing that would tarnish Gus's excellent service record was ever uncovered. In July of 1999, more than 32 years after his death in the Apollo launch pad fire, a search expedition financed by the Discovery Channel and spearheaded by veteran undersea salvager Curt Newport, found Liberty Bell 7 in waters 3,000 feet deeper than those which still embrace the Titanic. The spacecraft was taken apart, cleaned up and put back together and today you can view it at the Kansas Cosmosphere and Space Center in Hutchinson, Kansas. No, Gus was not a Kansan. He was born in Mitchell, Indiana. Gus was an Air Force Major at the time of his flight.

USS NOA (DD-841)

Back-up recovery ship for Friendship 7, the John Glenn mission of February 20, 1962

The Noa was a Gearing class destroyer and weighed in at a full displacement of 3,460 tons. She was just over 390 feet long, her GE turbines turned out 60,000 shaft horsepower, and she could exceed 36 knots. She carried a crew of 336.

A BRIEF HISTORY OF THE NOA (DD-841)

The second Noa was laid down at the Bath Iron Works in Bath, Maine, on March 26, 1945, and launched just over four months later on July 30. She was commissioned on November 2 of the same year.

(Navy photo)

With war having passed her by, Noa was off to the Mediterranean with calls at ports from Lisbon to Venice throughout 1946. She took part in fleet maneuvers in the South Atlantic in early 1947, and then spent the next several years undergoing overhaul and serving as a school training ship for the Fleet Sonar School in Key West, Florida.

She became a rescue destroyer for the escort carrier Mindoro (CVE-120) in the summer of 1949 and then into anti-sub training, part of a Hunter-Killer Group in Destroyer Squadron Eight and a second deployment to the Mediterranean stretching into early 1951. The next two years saw Noa in operational training along our East coast.

By November of 1953 Noa was back in the hunter-killer business as part of a group in Japanese waters, and then more of the same type of duty when rotated back to Norfolk and patrols in the Atlantic in the spring of 1954.

In 1956 she went back to the Med for a third time. It was time for a Norfolk overhaul in August 1957 and then a year or two of naval exercises in the Caribbean. By the summer of 1958, Noa took part in 6th Fleet operations during the Lebanon crisis, saw duty in the Persian Gulf and then headed back to Norfolk.

The Med again in early 1959 and again with the 6th Fleet. Then – back to the states and a new homeport at Mayport, Florida, operations off the Atlantic coast and the Caribbean and installation of the latest anti-submarine weapons at the Naval Shipyard in Philadelphia.

Noa completed her Fram I overhaul by May of 1961, rejoining the Atlantic Fleet and operated from Guantanamo, Cuba, for 6 weeks of refresher training.

Later that year she was off to the United Kingdom and combined exercises with the British navy in the Eastern Atlantic. After a month of that it was back to our side of the "big pond."

As 1962 rolled around, Noa had some more work done at Mayport and began briefings for her part in the upcoming John Glenn mission – our first man in orbit. When she sailed to become part of what was called Station 14, she was one of 24 ships that made up the recovery task force. As we will note frequently, the number of ships assigned to recovery in the early days of our space program was quite high. So much was unknown and had to be learned. With confidence and experience, that two-ocean fleet would be cautiously but steadily reduced.

Let the record show that the good ship Noa was a mere three miles from Colonel Glenn when he splashed down and the prime recovery ship Randolph wasn't. Noa and her crew enjoyed his company for some three hours before a chopper came along and took him back to the Randolph.

It was not to be the last time a destroyer would be called on to do the deed when a designated recovery carrier wasn't close enough. It would happen three more times.

Back to more regular naval assignments and two more deployments to the Mediterranean with the 6th Fleet taking Noa well into 1964. It was time for a regularly scheduled overhaul at Charlestown from September of that year through January 1965; and then off to the west coast of Africa as part of the Gemini 6 recovery force, but a malfunctioning target vehicle shut that mission down prematurely. Noa would have one more involvement with the space business as part of the large recovery group deployed for the Gemini 8 flight in mid-March, 1966.

There would be the arrival of new anti-submarine helicopters in January 1967, and then further work with the Fleet Sonar School at Key West in the next several weeks. She was back to the Mediterranean once again in June and operated there through November.

Noa ran out the rest of the decade with several overhauls at Atlantic coast facilities and began preparation for one more deployment to the Pacific.

Noa was decommissioned on October 31, 1973, sold to Spain in May of 1978, and scrapped in 1991.

THE FLIGHT OF MERCURY-ATLAS 6 (FRIENDSHIP 7) WITH ASTRONAUT JOHN GLENN ON FEBRUARY 20, 1962

We had already elevated the original 7 astronauts to near-cult status and with just two sub-orbital flights under our belts we were ready, as a nation, to idolize the first of that group to go where only two Soviet cosmonauts had gone before – into Earth orbit.

Whether John Glenn's selection as that man had anything at all to do with his straight-arrow reputation, as detailed much later in Tom Wolfe's "The Right Stuff" and other fine books to follow, Colonel Glenn turned out to be the right guy at the right time. Good looking, solid and serious, he was a casting director's dream.

Simply put, the flight of Glenn's Friendship 7, was designed to put a man into Earth orbit, observe his reactions to the space environment, and get him back to Earth safe and sound. All of that was accomplished; but, as would become the norm in just about all our space efforts,

not without anxious moments. Lift-off occurred at 9:47 a.m. from Launch Complex 14.

John Glenn orbited the Earth three times and suddenly we were asked to assimilate numbers that most folks, outside of those in the space program itself, were not really ready to cope with. Eighteen thousand miles an hour.... around the world in less than 90 minutes? You've got to be kidding. Well, make that 17,554 mph. Still...pretty darn fast.

Three times around this old Earth in a mere 4 hours, 55 minutes and 23 seconds at altitudes ranging from a high of 162 miles to down around 100. Our network anchors and their space specialist colleagues doled it all out in tantalizing detail and whet our appetites for more. Inevitably some media critics would go on to say that they were unduly dramatizing certain aspects of these missions, overdoing it on apparent dangers; but, in the long run, those anxious moments turned out to be totally valid and while they, indeed, kept us all glued to our television set, it was not without justification.

Approaching the end of his first orbit of the Earth, John Glenn was utilizing his attitude control thrusters (another new term to deal with), swinging the nose of the spacecraft back and forth. Suddenly, as he expected the nose to swing back, it wouldn't. Stuck. Now NASA's well thought out redundancy clicked in. Glenn simply switched off the offending automatic control system and went to a back-up system that did the same thing but almost immediately there was another problem.

This time it was the heat shield. A faulty indicator told Mission Control that the shield had been released prematurely, but this turned out to be a false indication. They would not find that out until the mission was over. So to protect themselves, and their precious astronaut passenger, they opted to leave the retro pack in place even after the retro rockets, which slowed the spacecraft for re-entry, had been fired. Normally the pack would have been jettisoned. This time it stayed to possibly hold the heat shield in place in case it really had been loosened.

None of the feared problems actually occurred but the listening and viewing public was kept fully informed and the drama had reached a new high. The ever curious public was getting more and more familiar with a guy named "Shorty" Powers, NASA's voice to the world and he was pulling no punches. Thanks to Powers' straightforward "telling it like it is" approach, the press was already telling us that Glenn's re-entry could well be, in Flight Director Chris Kraft's own words, "Short, fiery and fatal."

Because the chosen astronauts were all seasoned fighter and/or test pilots they repeatedly responded to cockpit crises with calm and professionalism and Glenn put all that into place on this day. As his spacecraft heated up to a glowing red and fragments of heat shield passed by his window, he calmly reported what he was seeing before plunging into the communications blackout that was ever to prove

nerve wracking for all concerned, from Mission Control to the nation's living rooms.

By the time he splashed down forty miles long and far beyond the prime recovery carrier Randolph, all the feared problems had been dealt with or simply hadn't existed. A well-rehearsed member of the recovery fleet, the destroyer Noa, DD-841, was only a few miles away. With Glenn remaining in the capsule, Noa hoisted America's first Earth orbiter aboard just 21 minutes after splashdown. He later helicoptered over to the Randolph. For the third mission in a row, no frogmen and no flotation collar. They were yet to come.

They lit up quite a few cigars in Mission Control and we may never know how many champagne corks were popped nationwide but the celebrations were many and joyous. America had itself a genuine hero. President John F. Kennedy, knowing where the action was, was quick to fly to the Cape where he greeted Glenn personally and took a tour of our spaceport. Within days the president hosted all seven of the astronauts at the White House, awarded John Glenn NASA's Distinguished Service Medal, and sent him off down Pennsylvania Avenue in a limousine with Vice-President Lyndon Johnson as Glenn was invited to address a joint session of congress.

Then…on to Manhattan to enjoy what that island does best – a ticker-tape parade and a medal from Mayor Robert F. Wagner.

Suddenly, NASA, which had only recently been under fire for (Good Heavens) not beating the Soviet Union in anything and everything, appeared forgiven. All was right with the world and our citizenry was patting itself on the back.

You would expect a spacecraft that was used to put the first American into Earth orbit would merit a special place in which to be preserved and you would be right. You will find Glenn's Friendship 7 (MA –7) prominently featured at the National Air and Space Museum in Washington, D.C. At the time of this flight, John Glenn was a Marine Lt. Colonel.

USS INTREPID (CVS-11)

The USS Intrepid was the prime recovery carrier for two manned space missions – Mercury-Atlas 7 with Scott Carpenter on May 24, 1962 and Gemini 3 with Virgil (Gus) Grissom and John W. Young on March 23, 1965.

Intrepid was an Essex class carrier of the so-called "short hull" group which meant 872 feet in length and a displacement of just over 27,000 tons when launched on April 26, 1943, but those figures would change through the course of several major overhauls undertaken from the mid 1950's into the early 60's. Her final numbers – 898 feet as converted with a beam of 103 feet, draft 32 feet and a new top weight of 33,292 tons.

(Navy photo)

A BRIEF HISTORY OF THE USS INTREPID (CVS-11)

The USS Intrepid was the fourth U.S. ship to bear that name, and her life began at the Newport News Shipbuilding and Drydock Company in Newport News, Virginia, sometime in 1942. She was launched on April 26, 1943, and commissioned on August 16 of that year.

Like many of her sister carriers, it was early 1944 before Intrepid would head for the Pacific and a war that was edging closer and closer to the Japanese mainland in what came to be regarded as the military's "island hopping" campaign. For Intrepid and several other carriers which became key players in one of the three "fast carrier" groups then operating, the first target was the Marshall Islands.

By the fall of 1944, Intrepid's aircraft were flying in support of our efforts to re-take the Philippines and the landing of ground forces on Leyte. She was heavily involved in the Battle for Leyte Gulf, which was actually four major actions and cost the Japanese four carriers and a destroyer. Japan's super-battleship, Yamato, was heavily damaged but not out of the war just yet.

It was during the vicious fighting around the Philippines in October and November that Intrepid was hit by several kamikaze attacks, losing 10 men on October 30 and 11 more on November 25. Despite these hits, the carrier never left her station in the task force but did turn and steam for San Francisco a day or so later for repairs.

She was back in action by mid-February of 1945 when the island war had brought our forces close enough to the Japanese mainland to enable our carriers, Intrepid among them, to launch powerful strikes on airfields on Kyushu. Desperate kamikaze action had become Japan's last-ditch effort to hold off its now certain defeat. While Intrepid supported the allied invasion of Okinawa in April, Intrepid took its last hit from one of the Emporer's suicide pilots. It cost the carrier 8 more of its men while leaving 21 wounded.

A day later Intrepid was headed back to San Francisco for more repairs but was back in the heat of battle by early August in time to send her planes to attack Japanese forces on the by-passed Wake Island. Just over a week later, the atomic bombs dropped so recently on Hiroshima and Nagasaki had brought World War II to an end and Intrepid was directed to "cease offensive operations" on August 15.

Between March of 1947 and March of 1962, Intrepid was in and out of the nation's shipyards becoming first, a modern attack aircraft carrier (CVA-11) and finally an anti-submarine warfare support carrier (CVS-11), which was the designation she carried into her retirement.

It was but a brief period after that last overhaul at the Norfolk Navy Yard that Intrepid was selected as the principal ship in the recovery team for Astronaut Scott Carpenter in Mercury Atlas 7 in May of 1962.

Most of the next several years would be devoted to her new role in anti-submarine warfare, but by March of 1965 Intrepid made her mark in the space program for the second and last time as prime recovery ship for NASA's first manned Gemini flight – Gemini 3.

Mid-1966 found Intrepid with the Pacific fleet off Vietnam.

Intrepid was decommissioned in 1974, assigned by Congress as the Bicentennial Exposition Ship at the Philadelphia Navy Yard in 1975-76, and finally to be used as a Sea-Air-Space museum in New York City in 1982 where she remains to this day, docked on the Hudson River just blocks from midtown Manhattan.

USS JOHN R. PIERCE (DD-753)

Back-up recovery ship for Astronaut Scott M. Carpenter and the Mercury-Atlas, Aurora 7 flight of May 24, 1962

(Navy photo)

A BRIEF HISTORY OF THE USS JOHN R. PIERCE – DD-753

This Sumner class destroyer began life at the Bethlehem Steel Company shipyard on Staten Island, New York, on March 24, 1944, was launched on September 1 of that year and commissioned on December 30 at the Brooklyn Navy Yard. The Sumner class represented the next evolutionary step from the Fletcher class and reflected the need for anti-aircraft warfare defense. They retained the same power plant but were given twin rudders and were slightly longer and wider in the beam.

After her sea trials she sailed for the Pacific on June 17 with her personnel doubtlessly aware that the war in that theater was fast winding down. On August 12, she was actually designated as part of a carrier-cruiser force sent to attack Wake Island but the A-bomb on Hiroshima and Nagasaki, Japan, put an end to that; and Pierce was told to stand down on August 15. For the next several months she played an important role in the movement of occupation forces and the liberation of Allied POW's. By late December she was helping to support the Chinese Nationalists in their struggle with the Communists for control of the mainland.

But her brief naval career was about to be put on hold along with dozens of other destroyers, carriers, cruisers and the like. Deactivation came in September of 1946, decommissioning the following January and entry into the reserve fleet in San Diego on May 1, 1947.

Returning to service in April of 1949, the Pierce became part of the Atlantic Fleet and took part in operations from Greenland to the Canal Zone right up until early 1951 when she was sent to the Mediterranean for a brief period. Then it was back to the Atlantic coast and operations

The John R. Pierce was an Allen M. Sumner class destroyer with a standard displacement of just over 2,600 tons* and a full load displacement of 3,218 tons. The ship measured 376 feet, 10 inches in length, had a 40 foot, 10 inch beam and a draft of 14 feet, 2 inches. She was powered by four Babcock & Wilcox boilers and could produce 60,000 shaft horsepower for a top speed of 36.5 knots. Her range was 3,300 nautical miles at 20 knots and she carried 20 officers and 325 enlisted men.

*We rarely found two sources on Naval history that agreed on all of the above specifications. One source we checked out for the Pierce had her weighing in a full 400 tons lighter, and they were a few knots apart on her top speed.

out of Norfolk before departing in May of 1952 for the Far East once again. The Korean war was in full swing and Pierce got into the thick of it with anti-mine, anti-junk and anti-fishing patrols and even took a few hits from enemy shore batteries in the process. But by December 12 she was back in Norfolk.

There followed six cruises of varying duration in the Mediterranean from early 1954 to April 1962. By this time, Pierce had logged no less than seven Mediterranean cruises but then it was time to play an important role in the space program. As but one of many ships deployed across two oceans for possible help in manned recoveries, the Pierce was part of the Atlantic Recovery Force's deployment for the Scott Carpenter mission, Mercury 7, and for the second time in but four such missions, a destroyer had to step in and do the job when the primary recovery carrier was too far from the scene at the time of splashdown. Overshoots would invariably lie with blips in the space mission itself and not the recovery fleet.

Actually another destroyer, the USS Farragut, was first on the scene but merely stood by. Meanwhile, helicopters from the Intrepid were able to approach and pick up astronaut Carpenter whereupon the Pierce arrived in the area and proceeded to safely retrieve the Mercury space capsule. She delivered the spacecraft to Roosevelt Roads, Puerto Rico, the next day, sailed back to Norfolk and then resumed duties in the Caribbean where she soon became part of the vast fleet, 90 ships in all, involved in the Cuban missile crisis later that year.

By March of 1963 Pierce was headed for maneuvers with the 6th Fleet in the Mediterranean and Middle East but was back in Norfolk by January of 1964. The Navy must have liked something about the destroyer Pierce and the Mediterranean because she was destined for one more deployment to that area in the fall of that year before returning to Norfolk and into duty as a reserve training ship – a mission which continued into 1967.

The John R. Pierce received one battle star for her service during the Korean War.

She was stricken from the Navel records on July 1, 1973. On November 6, 1974, she was sold and broken for scrap.

THE FLIGHT OF AURORA 7 (MERCURY-ATLAS 7) WITH ASTRONAUT M. SCOTT CARPENTER ON MAY 24, 1962

This one-man orbital mission came just about two months after John Glenn became the first American to orbit the Earth and Aurora 7 was, essentially, a duplicate of the Glenn mission. In fact, the length of the mission was less than a minute longer than its predecessor. To anyone believing such a small difference might not be of any consequence, consider that the Carpenter flight overshot the recovery carrier by 250 miles. More on that later.

The launch countdown on the morning of May 24, 1962, was near perfect. There was a 45-minute hold at T-11 minutes to tweak the camera coverage and for aircraft to check atmospheric conditions in the Cape Canaveral vicinity, but otherwise, all went according to plan. Because orbital missions required a lot more power than provided by the Redstone missiles, which put astronauts Shepard and Grissom into their sub-orbit flights, Carpenter was riding the bigger, more robust Atlas D, as did John Glenn. There were only minor differences between the two. Lift-off occurred at 7:45 a.m.

It was, by all measures available in this young program, an exceptionally good performance by both the Atlas launch vehicle and the spacecraft but, as usual, not without minor glitches here and there which were detected and dealt with.

Like John Glenn, Scott Carpenter was to orbit the Earth just three times in a further corroboration of man's ability to cope with orbital flight, weightlessness, G-forces and all the rest of the new experiences NASA was piling up data on.

When the time came to re-enter the Earth's atmosphere, Carpenter actually fired his retro-rockets three seconds too late. In his excellent new book, *For Spacious Skies*, Carpenter talks in detail about this incident. In his own words, "The rockets were supposed to fire automatically." But when that did not happen, Carpenter had to do the deed the old-fashioned way and push the retro-button himself. That is where those precious three seconds entered the picture and, coupled with an ensuing fuel shortage, resulted in Scott's landing so far beyond the primary recovery carrier Intrepid.

For all intents and purposes, Scott Carpenter was a missing person for almost 40 minutes before an Air Sea Rescue Service SA-16 amphibious aircraft reported a visual sighting. It was about that time Carpenter, sweating profusely in the 101 degree cabin, decided to get out of the spacecraft; but it was floating rather deeply and he feared the results if he opened the hatch. That was when he removed the control panel and squeezed out the throat, the narrow neck at the top of the spacecraft.

Then, about 36 minutes after splashdown, two aircraft circled the area and one of them, a Piper Apache, was taking pictures, so Carpenter knew he had been found. He had already inflated his own raft and was reasonably comfortable in it even as he waited for his rescuers. Twenty minutes later an SC-54 transport arrived on scene and dropped two Air Force pararescue men who quickly inflated two rafts and secured them to Carpenter's spacecraft and they struck up a conversation with a very relaxed astronaut who offered them some of his food and water. Another plane came over and dropped the flotation collar, which was also quickly installed. The destroyer USS Farragut became the first ship in the large recovery force to reach the capsule and it stood by. A short time later Carpenter was picked up by an HSS-2 helicopter from the carrier Intrepid, but now this episode got really interesting. NASA photographer, Gene Edmonds, kicked in with this part of the story. As the helicopter Gene was riding in neared the astronaut, hovered over him and lowered a sling to bring him up into the chopper, Gene said he had never seen so much green dye-marker on the water before. That dye makes visual sighting of a spacecraft from thousands of feet above a whole lot easier than it would be otherwise. It probably worked that way this time as well, but when the sling was lowered, it was given extra slack in case the helicopter rose any while Carpenter was putting it on. Scott was ready; he raised his head, gave a big thumbs-up signal and jumped off

the top of the capsule according to Edmonds. Two mistakes then took place. First, he should not have left the capsule until the sling line was tight — but worse, the hoist operator, in his excitement, pushed the down button instead of the up button when Scott jumped. Carpenter, observed Gene Edmonds, disappeared into the pool of slimy dye marker.

Edmonds was filming all of this with his Arriflex camera and reports that, in a few seconds, a bare green head popped up out of the Atlantic, followed by a green body. As he neared the helicopter in what is called a horsecollar, Gene saw green dye pouring out of every cavity in his space suit as they dragged him into the aircraft. He was coughing and choking and trying to hold onto his own camera. They cleaned him up in the helicopter on their way back to the Intrepid.

I had never heard this side of the recovery story before and it only came to me courtesy of NASA's Dr. John Stonesifer with Edmonds okay. There were more mishaps such as Carpenter's "Tale of the Green Dye" that were never given a whole lot of attention in the media. Probably because there were more important things to report on, and if the well being of the astronaut was fairly apparent that was good enough.

I could write another few pages just on gadgets such as urine bags which acted up or malfunctioned with very unpleasant results, which, in a few cases, caused everyone witness to the problem to have a good laugh — astronauts included.

But I digress. Scott Carpenter didn't look like the Jolly Green Giant when they got him back to the deck of the Intrepid and the destroyer USS John R. Pierce (DD-753), using its own special equipment moved in, hours later, to retrieve Carpenter's Aurora 7 spacecraft. As we noted earlier in this book, we give the John R. Pierce a nice assist in this mission.

At the time of this mission, Scott Carpenter was a Navy Lt. Commander.

Aurora 7 later found a home in Chicago at the Museum of Science and Industry.

USS KEARSARGE (CVS-33)

About midway through her naval life the Kearsarge became the prime recovery ship for two straight manned space missions – the last two Mercury flights; Wally Schirra in Sigma 7 on October 3, 1962, and L. Gordon Cooper's Faith 7 on May 6, 1963.

Kearsarge was one of "those". Meaning – Essex class on the drawing boards early on but subject to significant modifications and improvements which added to her original design length and created one of thirteen such carriers that would then be called Ticonderoga class. You will see this distinction noted five times in this book. In each of these instances...888 feet in length and a bow modified to accept two 40mm quadruple gun mountings for vastly improved air defenses.

(Navy photo)

A BRIEF HISTORY OF THE USS KEARSARGE (CVS-33)

Her keel must have been laid sometime in 1944 but she was not launched until May 5, 1945, as the war was winding down, and not commissioned until March 2 of the following year.

It would have been easy to forecast then that Kearsarge might spend her life without firing any shots in anger but few would have predicted years of fighting in Korea, tension over Taiwan, and our long involvement in Vietnam, but Kearsarge was engaged in all of those conflicts.

The late 40's consisted mainly of training operations and maneuvers leading to a mid-June 1950 modernization overhaul that would enable Kearsarge to handle new jet aircraft. By September of 1952, she was heavily involved in the Korean conflict as part of fast carrier Task Force 77 and in the next five months her planes flew nearly 6,000 sorties against Communist forces in North Korea. While serving in that theater her classification had been changed from CV-33 to CVA-33. She received two battle stars for Korean War service.

By 1958 Kearsarge had been overhauled, given the new angled flight deck and assumed her role in anti-submarine warfare service as CVS-33. The next several years saw Kearsarge kept busy with the 7th Fleet in the Far East.

The U.S. space program was moving into the headlines, and in 1962 and 63, Kearsarge had an interesting new mission.... prime recovery ship for the one-man Mercury flights of Wally Schirra and L. Gordon (Gordo) Cooper.

Kearsarge continued in her anti-submarine warfare role for years after that; but in 1970, made redundant by the general fleet draw down of the late 60's and early 70's, she was decommissioned. Three years in the Reserve Fleet followed. Kearsarge was stricken from the Naval Vessel Register in May 1973 and sold for scrapping in February 1974.

THE FLIGHT OF SIGMA 7 (MERCURY-ATLAS 8) WITH WALLY SCHIRRA ON OCTOBER 3, 1962

Our fifth manned space flight introduced us to Wally (never heard anyone call him Walter) Schirra. There was a "junior" after that name and I, as a network reporter, cannot recall anyone calling him junior either. From the beginning, he was "Wally."

The personality fit the name. Straightforward, tell-it-like-it is, easy-going and fun. The proof came when top network anchors like Walter Cronkite became extremely fond of having Schirra around for expert commentary and contribution during future space flights. Schirra was often outspoken and did not seem to worry about causing controversy.

So there he was on October 3, 1962, and after the usual built-in holds and checks and double checks, Schirra's Sigma 7 spacecraft lifted off from Launch Complex 14 at the Cape at just past 7:15 in the morning. The mission was planned for six orbits, just twice as many as the preceding two missions by John Glenn and Scott Carpenter.

Not only did Schirra hit the six but also later commented that he and the spacecraft could have gone on much longer.

There were good reasons for that optimism. The reaction control system, for instance, was modified to disarm the capsule's high-thrust jets during manual spacecraft operations and this saved precious fuel. Plus there was more "drift time" and that saved even more fuel. The record would show that any feared drift error was negligible.

Not only did the space program have its most talkative astronaut in

Earth orbit, they gave him two high-frequency antennas for improved communications; and to top that, he would take part in the first live television broadcast sent back to Earth during a manned U.S. flight.

This gave the space-hungry public its first real glimpse and grasp of life in outer space and NASA had the right man on duty for the task.

NASA reports total miles covered in orbit in nautical miles while other sources we have consulted use statute miles. For this book we will stick with the space agency and therefore Schirra's flight covered just less than 144,000 miles. Wally added a few more asterisks to his official flight record in becoming the first astronaut to splash down in the Pacific and only a few miles from the prime recovery carrier USS Kearsarge (CVS-33) about 275 miles northeast of Midway Island. Few of our manned space flights to come would ever get any closer to a recovery ship. Thanks to the efforts of the helicopter squadron HS-6 and the swimmers of UDT-11 and 12, Schirra was standing on Kearsarge's flight deck 37 minutes after splashdown.

That was October 3, 1962. On October 15, NASA Administrator James Webb awarded Wally the space agency's Distinguished Service Medal at a ceremony held in Schirra's hometown – Oradell in northeast New Jersey. Wally was a Navy Commander at the time.

Sigma 7, Schirra's Mercury spacecraft, can be viewed at the Astronaut Hall of Fame in Titusville, Florida. In other words, the Cape Canaveral neighborhood.

Given our fondness for nicknames, it was not surprising that press and public came to know this astronaut as "Gordo." He would seldom be referred to a Gordon…or L. Gordon just as Walter Schirra before him, was widely known as Wally.

This would turn out to be NASA's final Mercury flight, and its longest, although the space agency's decision to not continue with another Mercury launch would not be announced until nearly a month later. Actually it was a testament to Cooper's tenacity and skill since nearly every on-board system crashed at one time or another during Gordo's 22-orbit odyssey. He would frequently find himself turning to a manual operation when an automatic system let him down. That situation would continue right down to re-entry.

Lift-off occurred at just past 8:04 on the morning of May 15 after the usual glitches and holds had pushed it back a full day.

Thanks in large part to the fuel conservation measures developed on Mercury flights which preceded his, Cooper was able to give the space agency its first mission lasting more than one full day. Keep in mind, the longest U.S. space flights had been just six orbits. No wonder public interest reached a new high as Major Cooper sped to a new record for our space program with 22 orbits lasting one day…. 10 hours, 19 minutes and 49 seconds. But who's counting?

Gordo had a lot of time (when he wasn't combating on-board snafu's) to observe the world passing below him. At one point he actually deployed a small beacon satellite, which had strobe lights on it and was able to spot it next time around. You might recall that many cities and towns beneath the known flight paths burned extra energy to show themselves to the passing astronauts. There were efforts to "turn on every light you own" when space crews were known to be in the "neighborhood." In at least one instance, Cooper was able to spot a 44,000- watt xenon lamp in South Africa and know the town that it came from. He also got more sleep than any of his predecessors despite the difficulties he had to cope with.

As we noted earlier, Cooper had a lot of mishaps to handle during his flight. In today's parlance he would have been crowned the king of "cool" for his fine work. As it was, his automatic control system for re-entry was dying before his eyes, so he had no choice but to re-enter the Earth's atmosphere manually. This he did with near perfection and ended up landing just about as close to the recovery carrier USS Kearsarge as had Wally Schirra eight months earlier. In a repeat of their earlier fine efforts, the helicopters of HS-6 did the recovery job with the able assistance of the swimmers of UDT-11.

Like Wally Schirra before him, Cooper received NASA's Distinguished Service Medal soon after his voyage, but this time it was bestowed by President John F. Kennedy at the White House. At the time of this mission, L. Gordon Cooper was an Air Force Major.

Faith 7, as the last Mercury flight capsule was called, can be seen at the Johnson Space Center in Houston, Texas.

The Mercury one-man capsule was never designed to take our astronauts to the Moon and neither was Gemini, although there were some in the space program who proposed it as a means to attain circumlunar or lunar missions at a fraction of the cost and much earlier than Apollo. Those proposals were shot down because a ride in a Gemini space capsule was a hard lesson in living under extremely cramped conditions.

True, NASA now had a spacecraft which accommodated two of its astronauts but it actually offered only 50% more cabin space for twice as many people as had Mercury. Early on, the Gemini was christened (or nicknamed) the "Gusmobile" because the somewhat diminutive astronaut Gus Grissom was said to be the only astronaut who could fit into it. Gemini was 19 feet long, 10 feet in diameter and weighed about 8,400 pounds.

A crewmember, it was described, was crammed in, shoulder to shoulder with his partner and with his helmet scrunched up against the hatch. No stretching out. And without that luxury, some of the extended Gemini missions were said to be right up there in the painful category though we never heard any complaints from the astronauts themselves. If an astronaut stood up in his space after opening the hatch above his head and either stood there taking pictures or actually left the capsule on an EVA and ran into any sort of physical problem in the process, it was said that getting back into his seat and closing the hatch would have been very, very difficult. Fortunately, there is no record of such an event taking place.

(Courtesy of NASA)

The launch vehicles continued to come from the military. While Mercury capsules had been rocketed into space atop modified Redstone and Atlas missiles, the Gemini series debuted with Titan missiles powerful enough to meet the needs of a spacecraft, which now toted a Service Module (SM) immediately below it. Enter the modified Air Force Titan II. Once again it was a modified Air Force missile but this time manufactured by the Martin Company. A two-stage rocket 10 feet in diameter and 90 feet tall (109 feet with the spacecraft), burning Aerozine 50 and nitrogen tetroxide to produce 430,000 pounds of thrust in the first stage and 100,000 pounds of thrust in the second.

Now NASA had a two-man habitat designed to stay in Earth orbit for up to two weeks and powered by fuel cells located in the SM. As most NASA watchers will recall, that was when we became familiar with the process of jettisoning everything that was not the "Command Module" before attempting a fiery re-entry.

Gemini was created to teach our astronauts and Mission Control the vital skills of docking in orbit with another vehicle and of walking in space – EVA's or Extra Vehicular Activity. No lunar landings could be anticipated without them. There were roadblocks and setbacks, launch delays, and other anxious moments throughout Gemini but they were all lessons learned. By the time Gemini 12 completed its 59 Earth orbits on November 15, 1966, NASA could look to mountains of data and move on to project Apollo.

USS Intrepid was again the recovery ship

Much longer two-man Gemini flights were "just around the corner" so to speak, but for this first effort, NASA was content to check out and prove some basics and, consequently, it became a modest three-orbit mission. In the end it was rated highly successful but, as was becoming a habit, not without a few problems and one celebrated controversy. The wit and whimsy of astronaut Wally Schirra was about to make itself known. NASA hated it…. but the public "ate it up."

March 23, 1965. It had been nearly two years after the final Mercury one-man flight of L. Gordon Cooper and now this new two-man spacecraft that the crew named "Molly Brown" as in "unsinkable" was a nod to Grissom's ill-fated Liberty Bell 7 now sitting on the bottom of the Atlantic since his Mercury flight in July of 1961. NASA was less than thrilled with the whole idea but, in the end, decided it was better than the crew's alternative name – Titanic. The space agency really wanted to get away from the names attached to all the Mercury spacecraft but the Gemini's seemed to end up with well-known unofficial nicknames anyway. The astronauts made sure of that.

The lift-off at 9:24 that morning was perhaps the smoothest any of our astronauts had experienced up to that day. They would later tell reporters that it was less noisy than their moving-base simulator back in Dallas. Grissom was the veteran while John Young was the first astronaut not from the "Original 7" to leave a launch pad…. Launch Complex 19 to be exact.

Every flight produced one or more breakthroughs. They had to, in order to move the program ahead on schedule. One of the most important for Gemini 3 was the first use of OAMS, the Orbital Attitude Maneuvering System, which would create a controlled orbital and re-entry path. This meant the first fully maneuverable manned U.S. spacecraft.

Despite the relative brevity of their mission, the crew was able to alter the shape of their orbits by using on-board thrusters to practice and learn techniques that would be crucial in upcoming Gemini rendezvous and docking missions. Without docking…no Moon landing.

Now the controversy. The famed, or infamous (depending on your personal point of view) corned beef sandwich controversy. We have noted that Wally Schirra was one of the more fun-loving astronauts of the "Original 7" or any of those who followed in their footsteps. Wally made that distinction quite clear when he slipped off to Wolfie's Restaurant in Cocoa Beach, right near the Cape, ordered a nice corned beef sandwich and gave it to John Young who packed it away in his space suit. A no-no? Yes! Our astronauts were being trained to get their nourishment from toothpaste- like tubes and other packaging designed to keep their mealtimes from becoming a nuisance and an outright threat to their own safety. On this 3-orbit mission meals were not as much of a factor in the planning as they would be in all future flights. Young was authorized to eat some specially prepared "space food" while Grissom was not supposed to eat anything. Nonetheless when Gus got hungry, Young offered the sandwich to him and Gus ate only a portion of it, knowing the risks it presented. As it turned out…. that "part of a sandwich" still managed to send a lot of little crumbs into the cabin and could have caused major problems.

Needless to say, both men got thorough dressing downs from NASA's higher-ups and some of our lawmakers back in Washington had a few choice words to say on the matter. Only one of the men would fly again, with Young remaining in the program long enough to get to the Moon and become our first

Shuttle pilot; but the sandwich thing did bring stringent new regulations into play on what the astronauts could or could not bring with them on a mission.

While the press made a big deal out of a deli sandwich, the public, while not seeking to minimize the need for rules and regulations, got a decided kick out of the whole episode. Without a doubt it served to humanize these men who were still being depicted…basically…as above and beyond the ordinary.

During Mercury flights, the spacecraft would descend on one huge orange and white parachute and land, hopefully, upright, but not always. The flotation collars deployed by the frogmen who dropped from a carrier's helicopters were, therefore, circular, to encompass the base of a Mercury spacecraft. These new Gemini spacecraft would also be coming down under one big canopy, but the new system was designed to bring them in horizontally – on their sides in other words. Now the flotation collars took the shape of a Gemini as it appeared from top to bottom. You will find ample photographs of those collars in these pages.

Most of the time this new approach worked as designed, but…. not this first time. When first deployed, the big main parachute had Gemini 3 hanging vertically just like a Mercury spacecraft. Time to initiate the old "do it yourself" manual approach. The problem was when Gus Grissom hit the cabin switch the spacecraft pitched to horizontal instantly and the sudden drop sent both men into the spacecraft's

windshield. Grissom's faceplate was cracked and John Young's was scratched. No injuries.

To top things off, Grissom and Young found themselves some 60 miles from their recovery ship – the carrier USS Intrepid. Not the carrier's fault. Consequently, it took a little longer than it would on most missions for the frogmen to reach the space capsule; but they did, and quickly learned that Grissom had not forgotten his ordeal with Liberty Bell 7, so he would not crack a hatch door until those frogmen had safely attached the flotation collar. The swimmers were from UDT-21, which was beginning its mastery of Atlantic splashdowns. They would do them all with the exception of the last one – Apollo 9, and even that team (UDT-22) was from the same base at Little Creek, Va. The helicopter squadron was HS-3 out of Jacksonville Naval Air Station.

So Molly Brown did prove herself "unsinkable," but astronaut John Young was to echo Wally Schirra in saying later that the spacecraft "was no boat" as it bobbed and pitched on the Atlantic.

The bottom line on Gemini 3? All primary objectives achieved with the exception of the hoped-for controlled re-entry, which was only a partial success. A few other objectives fell into that partial category but, overall, Gemini got off to a great start.

Gemini 3 found a home in Grissom's home state. It is at the Grissom Memorial in Spring Mill State Park in Mitchell, Indiana.

John Young held the rank of Navy Lt. Commander at the time of this mission.

USS WASP (CVS-18)

(Navy photo)

The all-time manned recovery champion. Prime recovery ship for Gemini 4 (June 1965)... Gemini 6 and 7 (December 1965)... Gemini 9 (June 1966)... And Gemini 12 (November 1966). Earlier, the Wasp had been the back-up recovery carrier for the last Mercury one-man flight in May 1963.

She was Essex class.... pure and simple. Refer back to the Intrepid and the Mercury-Atlas 7 mission with astronaut Scott Carpenter (also Gemini 3) and you will find a ship of the same length – 872 feet with all other dimensions and statistics basically the same. Another term for these ships was Essex class – short hull.

A BRIEF HISTORY OF THE USS WASP (CVS-18)

There was something about a name like Wasp or Hornet that conveyed a deliberate threat, a warning to enemies of this country from the very beginnings of a U.S. Navy going back to the Revolution. You can be badly stung by a Wasp or a Hornet and that was always the message from the time of sailing ships to the two aircraft carriers which entered World War II, and the two ships which replaced them when each was lost to combat in the Pacific. The Wasp was the ninth to bear that feisty name.

When the keel for that Wasp was laid down at the Bethlehem Steel shipyard in Quincy, Mass., on March 18, 1942, her name was to have been Oriskany at launch; but building carriers takes time and by the time she was ready to slide down the ways and take her place in the U.S. Navy, the eighth Wasp had long since departed.

It happened on September 16, 1942, when Wasp (CV-7) and Hornet (CV-8) sailed in support of U.S. Marines on military transports heading for the long and bloody fight on Guadalcanal. Both carriers had done nobly in attacking Japanese ships and warding off the frequent attacks of Japanese aircraft but it was a Japanese submarine that got the Wasp. Shortly after 2 p.m. that day, three torpedoes slammed into the Wasp and it must have been clear to those on board and on nearby U.S. ships that the damage inflicted would be mortal. The eighth Wasp was abandoned after heavy damage and loss of life; and an American destroyer, firing 5 more torpedoes, sent her to the bottom sometime after 10 that evening.

When you look at the calendar then, you can see why what was to have been Oriskany in March of 1942 became the Wasp (CV-18) long before its launch in August of 1943 and commissioning on the 24th of November. In a strange twist of fate the same evolution surrounded the Hornet. Begun under another name but changed during her construction. Details later in this book.

Following the usual sea trials, and correction of the minor flaws which show up during those operations, the Wasp transited the Panama Canal, and, following training exercises in Hawaiian waters, took her place in mid May of 1944 as a part of the newly formed Task Group 58.6 which was part of Vice Admiral Marc Mitscher's Fast Carrier Task Force TF-58.

The U.S was prepping for the invasion of Saipan in the Marianas and Wasp's planes carried out raids on Japanese installations on both Marcus and Wake Island as part of that effort. They also attacked Japanese forces on Iwo Jima in June and flew in support of our landings on Guam on July 5.

By September Wasp was involved in operations leading up to the allied effort to take back the Philippines and, in which, the multi-day Battle for Leyte Gulf ultimately ended the Japanese fleet as a serious challenge to American supremacy in the far east.

Suffice it to say, you will find this period in the history of World War II dealt with in great detail in a library full of great books dealing with the war in the Pacific.

Wasp continued to play a vital role as our naval and ground forces moved up through the Pacific ever closer to Japan in the aptly named "island hopping" campaign designed by Admiral Chester A. Nimitz. By early 1945 Iwo Jima and Okinawa were next on his list, in that order, and they provided American bombers the access needed to carry out extended raids on the Japanese homeland.

Wasp took a bomb hit from a Japanese plane in March when kamikaze attacks were reaching their peak and her gunner's fired more than 10,000 rounds at the suicide-bent Japanese attackers.

By mid-April the Wasp was back in a shipyard in Bremerton, Washington, to get patched up and was headed back into the heat of battle by mid-July. Her planes conducted raids on the still Japanese-held Wake Island and hit nearly undefended targets in and around Tokyo as August was beginning.

It will be noted elsewhere in this book that many of the ships chronicled in this salute to their role in the space program saw their time in the Pacific abruptly halted by an awesome new weapon dropped on two Japanese cities – Hiroshima on August 6 and Nagasaki on August 9, 1945. Japan's leaders decided the game was up a few days later.

It must be noted here and now, before moving on with the Wasp story, that she received a total of eight battle stars for her World War II service. Commendable under any circumstances but all the more so for a ship that joined the conflict little more than a year before it ended. She proved that a ship named Wasp was not to be messed with.

But there was to be much more to the Wasp story. For a few years she functioned nicely as a troop transport but was sent to the reserve fleet in February of 1947.

In the summer of 1948 Wasp came out for refitting and alterations. It was time to deal with the larger, heavier, faster planes of the jet age.

She collided with a destroyer during night operations near Gibraltar on April 26, 1952. Tragically, the destroyer Hobson lost 176 of her crew. Wasp's personnel escaped without injury, but the ship's bow sustained a 75-foot saw-tooth rip that led to one of the strangest stories in U.S. naval annals.

We have spoken of the kinship shared by Wasp and Hornet and the histories they shared so proudly, not to mention the battles they took part in together. Now with repair badly needed on her battered bow, the Wasp pulled into drydock in Bayonne, New Jersey, while not far away in Brooklyn, New York, the Hornet was undergoing some modification of her own. Subsequently the Hornet's bow was removed, floated by barge down to Bayonne and fitted into position on Wasp. A remarkable task that was completed in ten days. Not unlike a family member sharing a kidney with another. Now the two ships were more alike than ever.

Into the fifties and a series of goodwill visits in the Mediterranean and then it was back to the familiar waters of the Pacific.

Wasp returned to San Diego in mid-October of 1956 where she was reclassified as CVS-18, an anti-submarine warfare aircraft carrier, and into the Boston Naval Shipyard just about a year later for a major overhaul (the angled deck among other changes) which took more than five months to complete.

Years of anti-submarine duty were to follow; but in November 1962, the Wasp became one of the many naval vessels deployed to enforce the U.S blockade of Cuba. It was a task that fell to a number of other ships also described in this book.

By mid-May of 1963 the Wasp was getting her first taste of America's budding space program when she took station off Bermuda as a back-up recovery ship for Major Gordon Cooper's 22-orbit mission aboard the Mercury capsule named Faith 7. It was only the beginning as you will see and read. Actually, the recovery was accomplished by the Kearsarge in the Pacific. Wasp was ready in case the mission had to be brought down in the Atlantic.

Over the next three years the Wasp would go on to serve as prime recovery ship for no less than five manned Gemini two-man missions and must have proved to CNO, the Chief of Naval Operations, that those assignments were richly deserved.

In between the five manned recoveries which Wasp performed so well from June 1965 to November 1966, the carrier would go right back to her regular mission in anti-submarine warfare exercises. During these assignments the Wasp called Boston her home port, but in June of 1968 that was changed to Quonset Point, Rhode Island, and would remain that way for the rest of the Wasp's naval service.

If you ever served on a ship such as the Wasp and were a crew-member at the time of decommissioning, it must be, for most people, a very emotional time. That sad news came on March 1, 1972. The actual ceremonies were held on July 1. She was sold on May 21, 1973, to Union Minerals and Allows Corp of New York City and subsequently scrapped.

USS Wasp, circa February 1967.

I only spent about a week aboard the Wasp in June of 1966 but even I can feel that sadness as I write about her. No small part of that has to do with the friendship I developed with her skipper, Captain Gordon Hartley. A friendship that continues to this day, more than 36 years later.

THE FLIGHT OF GEMINI 4 WITH ASTRONAUTS JIM MCDIVITT AND ED WHITE FROM JUNE 3-7, 1965

Up to this mission, the longest U.S. manned space flight had been Gordo Cooper's 22-orbit mission back on May 15 and 16, 1963. Now NASA was confident and ready enough to plan on nearly tripling Cooper's time in space. The first big step in testing men on long-duration missions is learning how one of those men would fare when stepping out of the spacecraft and going for "a walk" in outer space.

June 3, 1965. Lift-off occurred at 10:16 that morning and, beyond any doubt, the flight's primary objective was to test and evaluate the performance of the Gemini 4's systems for the mission's planned length of more than four days; but, of course, to learn at the same time, how two men would come through prolonged exposure to life in outer space. That was job number one.

Close behind, and not just because the Soviets had successfully put one of their cosmonauts, Aleksei Leonov, outside his Voshkod 2 spacecraft three months earlier, would be America's first "space walk," and that honor would go to Ed White.

Ed was the pilot while Jim McDivitt was the commander – the boss – so it was he who decided to delay the EVA (Extra Vehicular Activity) for one revolution beyond the book plan when other activities seemed to pile up. Mission control Houston, calling the shots from the ground for the first time, approved McDivitt's decision.

What had taken place was a futile effort to close on the jettisoned booster rocket for some valuable experience in "station keeping." The crew was new at it, brand new in fact, as was the team back in Mission Control. It never happened. Instead it used up a lot of precious fuel and left Ed White looking tired to Commander McDivitt and thus – the one orbit delay.

When the time came for Ed White to step out of Gemini 4, it was halfway through the spacecraft's second revolution and White would remain tethered to Gemini 4 by NASA's so-called golden "umbilical cord" which provided his life support and communications with McDivitt and Mission Control. He was using a hand-held jet thruster to move around in space. We didn't have TV from space at that time, but Jim McDivitt was taking both spectacular color stills (they had a 70 mm Hasselblad with them) and also 16 mm film, which we were to see soon after the mission ended. The Soviets showed their people nothing and only reported on their space efforts weeks after the fact.

It was a special time. The U.S. was moving up on the Russians in flight durations and outdoing them in some key, dramatic areas such as walking in space.

Next we were told that Gemini 4 would go into "drifting mode" for the next two and a half days. It was not too difficult for us landlubbers down here to understand what this meant in fuel savings even though few of us had any idea what those fuels were. NASA's onrushing new terms and technology constituted real information overload in many areas.

The mission rolled on and because of its duration, the crew found themselves using a bungee exerciser more than they thought they would and recommending a systematic exercise program as vital when the missions got even longer.

When a key computer failed, the astronauts had to give up on the planned computer-aided re-entry and revert to what was called a rolling Mercury-type re-entry; but these adjustments were almost

getting to be routine, if not terribly welcome. They were giving NASA and its astronauts incredible experience on what to expect in space and how to deal with challenges from A to Z.

Time to head for home. Sixty-two impressive orbits of the Earth and the aircraft carrier USS Wasp waited in the Atlantic. Once again a close-to-the-prime recovery ship splashdown was not in the cards and Gemini 4 landed just over 50 miles from the Wasp. Wasp had played back-up recovery ship for the last Mercury mission and this would be her first of five assignments as primary recovery ship in the Gemini program. No other ship would touch that fine record.

Despite missing its mark, swimmers from the Wasp (UDT-21) arrived over the spacecraft within minutes in one of the HS-11 helicopters. McDivitt and White were lifted into the chopper quickly and were standing on Wasp's deck less than an hour after their splashdown.

As for the number of ships deployed, trained and ready to perform a recovery in either the Atlantic or the Pacific, Gemini 4 turned out to be almost a high water mark in that area, as 26 ships were on the job where they could be called upon if needed. Only Gemini 3's 27 ships exceeded that compliment. But it was the mission with the most aircraft assigned at 134, and that included Wasp's helicopters and fixed-wing aircraft along with planes ready and waiting on Bermuda and numerous other participating bases around the globe and "under" Gemini 4's orbital flight path.

McDivitt and White had paved the way for longer missions to come. President Lyndon Johnson came to Houston to congratulate and honor them. About a million Chicagoans showed them how the Windy City could put on a ticker-tape parade, and, at the President's request, they went to the Paris International Air Show where they met with Yuri Gagarin, the world's very first man in space.

Gemini 4 ended up in some very distinguished company. You will find it along with a lot of other great NASA hardware at the National Air and Space Museum in our nation's capitol.

Jim McDivitt was an Air Force Colonel at the time of this mission. Ed White was an Air Force Lt. Colonel.

USS DUPONT (DD-941)

(Navy photo)

A BRIEF HISTORY OF THE USS DUPONT – DD-941

This may be the briefest history to appear in this book. We have appealed to a host of web sources, naval history sites and so forth. Very little information has come back to us. We have learned from several individuals that some ships simply did not go to the trouble to chronicle their service histories to the degree that so many others have, and that is unfortunate.

We have endeavored to pay tribute to all these ships, to their wartime or peacetime records, to their special achievements and awards, but in the case of the good ship DuPont the cupboard is almost bare.

Besides the data on the previous page we do know that she was built by the Bath Iron Works in Bath, Maine, launched on September 8, 1956, and commissioned on July 1, 1957. She had an interesting naval life in the brief notes we were able to find. Her very first month in service found her serving on a midshipman cruise and anti-submarine exercises in the Atlantic ending with a visit to New York. In September DuPont sailed for the Mediterranean and a tour of duty with the 6th Fleet and took part in highly realistic air defense and anti-submarine warfare problems. She headed back to Norfolk in March of 1959 to get ready for Operation Inland Sea, the historic first passage of a naval task force into the Great Lakes through the Saint Lawrence Seaway. In that passage she escorted HMS Brittania with Queen Elizabeth II of England during dedication ceremonies in late June.

DuPont crossed the Atlantic again in September, visiting Southhampton, England, after serving as a plane guard for the trans-

The destroyer DuPont was back-up to the USS Lake Champlain on Gemini 5 and did the actual astronaut recovery. DuPont was a Forrest Sherman class destroyer, weighing in at 4,619 tons fully loaded. She was 418 feet long with a beam of 44 feet and a draught of just under 12 feet. Her power plant turned out 70,000 shaft horsepower fed to two screws and she could attain a speed of 33 knots with a range of 4,500 miles at 20 knots.

Atlantic flight of President Dwight Eisenhower. On the following January 28, Dupont again departed Norfolk and took up a second tour of duty with the 6th Fleet in the Med, returning to Norfolk at the end of August for an overhaul in the Naval Shipyard there. She remained there through the end of the year.

This is where the mystery begins with the end of 1960. The preceding information is based on research available from various web sites. Yet the record is quite clear insofar as NASA records are concerned. Obviously the DuPont was part of the recovery forces arrayed across the Atlantic when Gemini 5 splashed down more than 90 miles from the prime recovery ship, the Lake Champlain, and, as

detailed on the next several pages, Navy divers from the DuPont did the job of rescuing the crew and taking them by helicopter back to the Lake Champlain, at least according to a NASA website. That was in August of 1965, but none of the naval records we checked went that far. Far from it.

At that point, our search for further DuPont history failed to produce anything except a date for her decommissioning on March 4, 1983, stricken from the Navy's records in June of 1990 and scrapped in 1993. That leaves over two decades largely absent and unaccounted for….AWOL.

THE FLIGHT OF GEMINI 5 WITH ASTRONAUTS L. GORDON COOPER AND CHARLES "PETE" CONRAD FROM AUGUST 21 TO 29, 1965

Recovery ship: USS DuPont and/or USS Lake Champlain

Will the Real Recovery Ship Please Stand Up!

As you might expect, the third flight in the two-man Gemini series would go on to be the longest flight yet for any American astronauts. The program had to grow, had to prove man's ability to function and survive in space for longer and longer periods of time.

Consequently, this 8-day excursion would not only show NASA that men could stand up to space-flight time sufficient to get them to the Moon and back, it also tested the nerves and patience of those men. When Gemini 5 and its considerable "drifting" time in space to conserve fuel was over, Pete Conrad would comment that it was just about the hardest thing he had ever done. The word "boring" was injected frequently into his post-flight press briefings.

Conrad and Cooper lifted off from Launch Complex 19 at the Cape at a blink of an eye before 9 a.m. on August 21, 1965. Their spacecraft and its launch vehicle combined were called Gemini-Titan II GLV-5; and

while the crew had their own name for their little home, there was no official capsule nickname.

In what had now become routine and almost expected, the flight had been pushed back from August 19 due to bad weather in the area. Despite Pete Conrad's description of their mission as basically boring, they did have 17 experiments on the books and were able to carry out all but one of them, and that was only because they could not photograph a REP, Rendezvous Evaluation Pod, which they had launched earlier in their mission. That became impossible when the fuel cell system acted up. The fuel cell was designed and built to provide both water and electricity during all future missions; and so, obviously, would have to work properly in order for any consideration to be given to flying to the Moon and back.

There were many other functions that did not come under the heading of "experiments" which brought their own problems to the crew. The mission was barely into its third orbit when they noticed that pressure in the oxygen supply tank of the fuel cell system was falling, but the matter was stabilized and thereafter improved during the remainder of the mission.

On their fifth day in space, the OAMS (Orbital Attitude Maneuvering System) became sluggish and several thrusters became inoperative. This curtailed any number of planned operational activities.

Whenever Gemini 5 went into its fuel-saving "drift" mode, the cabin would get cold. The coolant circuit in their space suits also seemed too cold; so, at one point, they actually took the hoses off to stop the coolant from circulating. They had trouble sleeping and tried to sleep alternately, but calls from ground control kept waking them up; so they reverted back to sleeping, or at least trying to sleep simultaneously, and this was only partially successful.

Since there was no docking vehicle with them in space as yet, the two men did carry out a series of maneuvers with an imaginary or "phantom" Agena unmanned tracking vehicle. Agena's would be launched in future flights and became the indispensable training device for rendezvous and docking.

The fuel cell began producing too much water and the spacecraft had no system for getting rid of it. Ground control figured out, that despite the overproduction, there would be still be room for the water at the end of the mission. Picture a big sigh of relief in space…and on the ground.

Now some human error on the ground crept into the mission as it neared time for its re-entry. Hope you can follow this. Earth's rotation (so they tell us) is 360.98 degrees per day. Got that? Well, in programming the computer, someone left off those two numbers behind the decimal and fed the machine a straight 360 degrees. Command Pilot Gordon Cooper recognized what he thought was an erroneous reading and he was correct in that assumption. As it turned out, his efforts to compensate for that error actually brought them down closer to the recovery carrier, USS Lake Champlain, than they might have been. Even so, they dunked into the Atlantic about 90 miles short.

And This is Where the Controversy Begins

There were 19 ships scattered across two oceans for this mission and one of them, the destroyer USS DuPont (DD-941), was a lot closer than the Lake Champlain.

This is an exact reprint of a page from a NASA website. This same wording also appears in a NASA science website.

Launch	Landing
August 21, 1965; 8:59:59.518am EST. A launch attempt on August 19 was postponed due to weather conditions and problems with loading cryogenic fuel for the fuel cell.	August 29, 1965. Landing was at 29deg44min North and 69deg 45min West. Miss distance was 170.3km (92nm). Navy divers from the backup recovery ship USS DuPont (DD-941) recovered the crew and transferred them via helicopter to the USS Lake Champlain (crew onboard in 89 min).

However, on the Navy website, "A Brief History of Aircraft Carriers," it simply says, "Navy frogmen helped astronauts Gordon Cooper and Charles Conrad out of their space capsule and aboard a helicopter for the ride back to Lake Champlain." It does not say exactly what frogmen nor where they came from. It also does not tell us where the helicopter was from – which squadron?

Regardless of how you consider the above, it is a fact that most sources name the Lake Champlain as the recovery ship. Are they only stating the fact that Champlain was the designated recovery ship and not necessarily the one that actually did the job? We feel there is significant room for doubt.

Nonetheless, although it was a somewhat troubled mission, NASA could now be confident that its astronauts could stand whatever space had to dish out on a mission to the Moon and back. NASA doctors were amazed and delighted at the physical condition of the Gemini 5 crew. Heartbeat rates, for instance, had returned to normal by their second day back from space.

President Lyndon Johnson sent the men on a six-nation goodwill tour, and while they were on that trip, they attended the International Astronautical Federation Congress in Athens, Greece, and met another two-man space crew, Soviet cosmonauts Aleksy Leonov and Pavel Belyayev of Voshkod II.

Pete Conrad was a Navy Lt. Commander at the time of Gemini 5.

USS Wasp was again the recovery ship

The crew of Gemini 6 was Wally Schirra and Tom Stafford while the Gemini 7 crew was Frank Borman and Jim Lovell. For reasons we will explain very shortly, Gemini 7 took to flight on December 4 and remained in space until December 18, 1965, while Gemini 6 followed on December 15 and was back on Earth (in the Atlantic anyway) two days ahead of Gemini 7. Sound a bit confusing? It was.

As originally planned, Gemini 6 was supposed to go first and rendezvous and dock with the unmanned Atlas-Agena target vehicle – Atlas being the rocket which carried it into space and which had previously been used to power the last four Mercury one-man flights but was not powerful enough to propel a two-man Gemini.

But the best-laid plans do not always work, do they? With Gemini 6 and its crew already sitting on the pad at Launch Complex 19, the Agena blew up shortly after lift-off. The Gemini 6 launch was immediately shut down. Now what to do? A few imaginative and daring officials of McDonnell-Douglas, builders of the Gemini, were quick to think about using the manned Gemini 6 as a target vehicle for the upcoming Gemini 7. There would, of course, be no docking as neither spacecraft had the docking collar that Agena had been equipped with for that very specialized maneuver. But how about simply maneuvering as close as would be possible and safe? Most NASA officials thought the idea was outrageous–impossible. Not enough time and too risky.

There were good reasons for their skepticism. Given the maximum duration in space of the Gemini fleet at that moment, how could NASA possibly launch two missions within the known time frame of about two weeks? You were looking at a launch pad turn-around time between launches that had never ever been discussed before. Those pads sustain a certain amount of damage after every launch from the intense heat and power of a lift-off. Could crews get in there, make the pad safe and secure enough in such a short time to allow another mission in under the prescribed time?

Well, those McDonnell-Douglas officials, Walter Burke and John Yardley, didn't seem overly concerned about it and the four astronauts involved, Schirra, Stafford, Borman and Lovell all spoke in favor of the idea; but it was, of course, not their decision to make. Just four gung-ho fly boys ready to go for it.

The saga of trying to sell the idea of two manned launches within two weeks went back and forth at NASA's top levels and even reached into the White House before it was decided to give it a try. On Thursday, October 28, at a news conference at LBJ's Texas White House, the new Gemini 7/6 mission was officially announced. Presidential press secretary, Jim Moyers, spoke of a projected start date for January; but, meanwhile, back at the Manned Spacecraft Center, everybody there was pointing toward something more like early December, and that is exactly how it would turn out.

Nothing tricky…just a nice close rendezvous. Stafford and Schirra, definitely the more adventurous of the two crews, even talked about one member of each crew trading places in space, but Frank Borman showed his caution and made it clear that he would have no part in opening a spacecraft hatch out there. It might look great in the headlines, said the future head of Eastern Airlines, but one little slip and, in his words, "You have lost the farm."

So it came to pass that Gemini 7, which had been planned as a long duration space flight anyway, would launch ahead of Gemini 6 which would then head into space, catch up to Gemini 7, rendezvous, chit-chat and quickly return to Earth.

When it came time to put this whole risky business into motion, the capsule that had been Gemini 6 was now called 6-A in a sort of recognition of the late, lamented Agena target vehicle.

Now you know (thank you Paul Harvey) the rest of the story. Gemini 7 streaked into the heavens at 2:30 (EST) on the afternoon of December 4, 1965. It would be eleven days before the newly planned rendezvous with Gemini 6 would take place, and by that time, NASA had itself yet another standard for man's ability to survive and function in space, surpassing Gemini 5's previous record of 8 days. There was a whole lot more to this mission than simply going into orbit and waiting for Gemini 6. As was becoming more and more routine, the men were extensively attached to all sorts of medical monitoring devices that told Mission Control how they were dealing with the confines of their spacecraft, extended weightlessness, their special diets and much more.

Before launch, Frank Borman and Jim Lovell, learning from their predecessors, had decided that their sleep and work times would coincide. They even had to look around that small world they would call home for two weeks for new places to store their waste paper from those new-fangled meals packaged in tubes and envelopes. The solution was behind their seats with each astronaut's seat eventually storing the waste paper for seven days.

They would also wind up wearing a newly-developed lightweight pressure garment to allow them to get out of those bulky, cumbersome space suits the world had become familiar with. They had a margin of safety built in just in case their regular space suits would suddenly be needed.

They would carry out, or try to, no less than 20 experiments, eight of which were medical. There would be calcium balance studies, in-flight sleep analysis, and even brain studies (EEG's) which the astronauts were less than thrilled about. They had to monitor food intake and body wastes not just for 9 days before the flight, but the 14 days during it, and 4 days afterward. To this day, even in the space shuttles, the question of "how do they go to the bathroom" in space remains the one area that the crews least like to talk about. This, despite the fact, that the process is much more acceptable to both men and women in space than it has ever been.

December 15 arrived and a mood of high anticipation was easily detected among planners, pilots and ground crew. At 8:37 in the morning, Gemini 6, with Schirra urging the booster's engines on, roared to life and headed for its historic meeting with Gemini 7.

Already, Soviet space endurance records had been shattered on two consecutive flights and now this. The Soviet space agency was bragging about a space rendezvous but that turned out to be a distant meeting with two Soviet spacecraft not even in the same orbital plane. This effort was to end with two Gemini's within mere yards of each other and they would come back with the pictures to prove it.

Now the skills of Wally Schirra and Tom Stafford were in evidence as the two men, but mainly Schirra, fine-tuned the distances separating the two Gemini's on every orbit. Closer and closer until they were joyously waving flags back in Mission Control, lighting those cigars and smiling at each other to beat the band. The two space ships would fly together with distances ranging from a few hundred feet at times, down to as close as 20 feet. This was thruster time, using short bursts to dance around Gemini 7 and certainly close enough that the two crews could wave clearly at each other as well as converse over their radios.

At one point, Frank Borman in Gemini 7 remarked to a tracking ship communicator, "We have company tonight," as bedtime (if we can really call it that) approached for both crews. After their guests had departed and Borman and Lovell realized that all the excitement and anticipation of rendezvous was over, the mission began to drag. They still had three days to go and Jim Lovell got around to musing about his legs, how they were nearly useless in space and "just there."

Now the Gemini 7 crew began to have fuel cell problems and a warning light threatened to shorten the planned 14-day mission that the astronauts earnestly wanted to achieve despite their cramped quarters. Test results back on Earth showed that their electrical system would, indeed, take them the full length of the mission, and so it did.

They then proceeded to carry out NASA's first controlled re-entry. On prior missions, a crewmember had to step in at one point and activate systems by hand, so to speak. Now the computer was calling all the shots and it did...four times.

Borman let Mission Control know that after splashdown the two of them wanted "out of there" as quickly as possible and would not ride the spacecraft back to the deck of the carrier Wasp. They missed the recovery ship by about six miles, but frogmen from the Wasp (UDT-21) were on the scene in short order, affixing the flotation collar and assisting, as they were "horse collared" on a cable up into a recovery helicopter (HS-11). Wasp had carried out its second successful recovery in just two days

(Gemini 6 on December 16). The men were a bit gimpy-legged when they first hit the deck, but NASA doctors were delighted to see them both spring back to normal, or very close to it, after some rest, cleaning up and real food. Their post-flight tests went very well.

NASA had its best Christmas present yet. Ample evidence that man was ready for Moon mission duration, rendezvous in space and controlled re-entry. Santa Claus delivered.

More good stuff for the National Air and Space Museum in Washington, D. C. That's where you will find Gemini 7 while Gemini 6 ended up at the Science Center in St. Louis, Missouri.

At the time of this mission, Tom Stafford was an Air Force Colonel as was Frank Borman. Jim Lovell was a Navy Commander.

USS LEONARD F. MASON (DD-852)

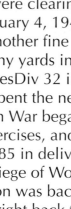

The destroyer Leonard F. Mason became the recovery ship for the Gemini 8 mission (March 16 and 17, 1966) with Neil Armstrong and David Scott, when that mission had to be aborted and suddenly brought down in the Pacific. The carrier USS Boxer was the designated prime recovery ship in the Atlantic.

The Mason was a Gearing class destroyer. Her statistics, therefore, are nearly identical to those listed for the USS Noa (the John Glenn mission – Friendship 7).

A BRIEF HISTORY OF THE DESTROYER LEONARD F. MASON (DD-852)

This Gearing class destroyer was starting to grow from its keel upwards May 2, 1945, when the clouds of World War II were clearing in both Europe and the Pacific. She was launched on January 4, 1946, and commissioned on June 28 of that year. Another fine product from the Bethlehem Steel Company yards in Quincy, Mass. Her first deployment was with DesDiv 32 in the Pacific on January 22, 1947, and she spent the next several years in that theater. As the Korean War began the Mason joined in anti-submarine exercises, and on May 16, 1950 she joined Task Force 85 in delivering continuous shore bombardment at the siege of Wonsan. By August of 1951 the Leonard F. Mason was back in San Diego for an overhaul, but then went right back to the coast of Korea and Wonsan Harbor. Except for another trip back to California in late 1952 she would return once again to Korean shores and serve there until the close of hostilities.

She departed Yokosuka, Japan, for Long Beach, California, in December of 1953 and was readied for peacetime duty. Between 1954 and 1960 there were three more WestPac cruises and duty with Fast Carrier Task Force 11 during the Suez crisis. Yokosuka was her homeport for anti-submarine patrols and other peacekeeping missions from May 1960 to May of 1962. During 1963 the Leonard F. Mason underwent FRAM I conversion at the Boston Naval Shipyard before returning to Yokosuka and two years of serving with various task groups of the 7th Fleet. This deployment included gunfire

(Navy photo)

support missions off the coast of Vietnam, patrolling the Taiwan Straits and – serving in the Gemini Recovery Force. It was at this point in her naval life she showed just why the recovery forces, especially in the earliest years of manned space flight, numbered in the dozens of ships in a number of oceans. Leonard F. Mason became very important to NASA and astronauts Neil Armstrong and David Scott aboard Gemini 8 on March 17, 1966. More details on this part of her history to follow.

Beyond that surprise assignment she went right back to her "regular" job of gunfire support off the coast of Vietnam. More trips back to California for overhauls and more duty back in waters off Vietnam. Leonard F. Mason received three battle stars for Korean service. The Leonard F. Mason was stricken from Naval records on November 2, 1976, but was still active in the Taiwanese Navy as recently as May 1998.

THE FLIGHT OF GEMINI 8 WITH ASTRONAUTS NEIL ARMSTRONG AND DAVID SCOTT, MARCH 16-17, 1966

This mission, though woefully short and with a life-threatening reason, was one for the books. It showed beyond doubt why the fleet of potential recovery ships was as big as it was and just how well trained and ready these support vessels were.

Gemini 8 was scheduled to be just a 3-day mission with only a handful of experiments planned. As it turned out, only one of that "handful" could be completed; but that one operation was, nonetheless, very important.

NASA still had not recorded its first successful docking with another vehicle in space. Remember 7 and 6 got very close to each other but could not dock because neither spacecraft was equipped to do it.

What was fitted with the proper collar for docking, was the Agena Target Vehicle, and that had exploded just before Gemini 6 was supposed to take off. Well, we went through that incredible episode just a few pages back; so let's move on.

Roll out another Agena; ready it for launch on pad 14 at the Cape, and send that unmanned hardware into space at 10 a.m. on the morning of March 16, 1966. Then, less than two hours later, send Gemini 8 up to catch it and dock with it.

Things got off to a promising start with both elements of the mission going into space with no noticeable problems. Just 6 hours and 33 minutes after Gemini 8 had been launched, it had zeroed in on the target vehicle. Command pilot Neil Armstrong was doing an expert job of firing his thrusters to alter their orbital path and their speed. He

could not know then that the thrusters that were working so well in getting he and Dave Scott "kissing close" to the Agena would quickly malfunction and send them both on a hair-raising ride.

It was the OAMS, (Orbital Attitude Maneuvering System) so vital to any mission, which now acted up. Shortly after docking, the spacecraft and the Agena went into a totally unexpected roll and yaw motion. Roll is just what it sounds like. The spacecraft was tumbling at a rapid and sickening rate. Yaw can be pictured as an object moving back and forth and from side to side. Now both were happening at a wild rate, and because both Mission Control and the astronauts in Gemini 8 believed the problem was in the Agena target vehicle, the decision was made to undock. They didn't know it then, but the OAMS had a short circuit and would work one minute and not work the next.

Armstrong would regain control momentarily and then the problem would resurface. They undocked and lo and behold, the spacecraft started spinning even faster – one revolution per second. Think seasick…air sick…very sick. Eventually the crew was able to stabilize their home in orbit but they used up 75% of their precious fuel in the process. Despite their sickening experience, both astronauts wanted to keep on and especially carry out Dave Scott's planned space walk, but Mission Control said, "No dice fellas…you're coming home."

Meanwhile, back in the Atlantic, the expected primary recovery carrier USS Boxer had little reason to believe it would not be carrying out its first recovery assignment in a few more days, and only learned later that Gemini 8 would be making a hasty return to Earth…but in the Pacific.

By now, the recovery forces had been winnowed down from a high of 27 ships on Gemini 3 the year before, to 16, as had been the case with the combined Gemini 7 and 6 missions. One of those 16 was a destroyer named Leonard F. Mason (DD-852). The Mason had done the drill, done its homework, and had rehearsed recoveries with a boiler-plate, so called "dummy," spacecraft many times, as had all other ships destined to be assigned as back-ups and reserves. Because it did not carry aircraft as the carriers did, it could not rapidly deploy a team or teams of frogmen in helicopters, as was standard practice aboard the prime recovery carriers.

That job was handled by two Air Force pararescue planes (HC-54's) flying from Okinawa and Japan. The plane from Okinawa got there first and in time to see the spacecraft moments before the splashdown. A team of three PJs parachuted down. The Air Force version of frogmen were able to stabilize the spacecraft, even in a rough sea, but only that onrushing destroyer could retrieve the capsule and crew, and it was still hours away when it was first told it would have to do the job.

The Mason gradually worked its way up to a top speed of 32 knots, but it was still a long wait for the Gemini 8 crew. As astronaut Wally Schirra was to say at a later date, "It's a great capsule…but a lousy boat." During its race to the splashdown point, the ship's staff doctor, Lt. Paul Fukuda, went over his instructions for doing a medical de-brief on the astronauts when they came aboard. Obviously, the Mason did get there and was quick and efficient in getting Armstrong and Scott on board.

Since this recovery was different from most, we will take a few lines to look at what took place once the crew was on the destroyer. Asked what they might like to eat, they both said, "Whatever was on the menu." That turned out to be spaghetti and meatballs with a shrimp salad added on. Later, at midnight, they happily put away some steak and eggs. The space duo toured the ship and received numerous souvenirs while giving out a whole lot of autographs. The Mason pulled into Okinawa the next day where fellow astronaut Wally Schirra was waiting to greet them. A short time later, they were headed back to Houston by plane.

In its short life in space, Gemini 8 did complete the program's first hard docking, but, as we have seen, it was not to last long. Both Armstrong and Scott would not only fly again, but each would have their chance to walk on the Moon in separate missions – Armstrong on Apollo 11 and Scott on Apollo 15.

Here is a NASA photo that tells the tale better than mere words. You are looking at the Mason's hoist as it lifted Gemini 8, with astronauts Armstrong and Scott still inside, aboard the ship. The flotation collar was still attached, but note the hard triangular "collar" right above it. Any destroyer assigned as part of a recovery fleet, Atlantic or Pacific, was equipped with such a collar and trained in how to use it.

Guess where Neil (first man on the Moon) Armstrong was born? The fact that Gemini 8 can be viewed at the Neil Armstrong Air and Space Museum in Wapakoneta, Ohio, may give you a clue.

Neil Armstrong became the first civilian to fly in the space program, but he would not be the last. Prior to his entry in the astronaut corps, Armstrong had been one of our most experienced test pilots and was a Naval aviator before that, having flown many combat missions in Korea.

David Scott was an Air Force Colonel at the time of Gemini 8.

USS BOXER (CVS-21 AND LPH-4)

A BRIEF HISTORY OF THE USS BOXER (CVS-21 AND LPH-4)

This fine ship qualifies for an Honorable Mention in the context of this book. She was in the right place at the wrong time. Boxer had the assignment as primary recovery ship for Gemini 8 and was right where she belonged when things went terribly wrong aboard that spacecraft and it had to be brought down hurriedly but in the South Pacific. So we pay tribute to the Boxer anyway. She was ready to do the job.

Boxer was built by the Newport News Shipbuilding and Drydock Company of Newport News, Virginia. Launched on December 14, 1944, and commissioned on April 16, 1945. With the demands on initial training and sea trials there was no way the Boxer would see duty in World War II, but from September of 1945 to 1957, she made no less than ten deployments to the western Pacific, a period which embraced the Korean war years and saw Boxer not just carrying war planes to that zone for both the Air Force and the Navy but also providing air support to the forces fighting ashore.

By the time she made her final western Pacific tour in '56 and '57, she had been converted to an anti-submarine warfare carrier (CVS).

As 1957 was nearing an end, Boxer operated briefly as an experimental assault helicopter carrier. When 1958 rolled around, she was transferred to the Atlantic fleet as an "interim amphibious assault ship" and received her final designation as LPH-4 on January 30, 1959.

For the next decade, Boxer and her compliment of combat-ready Marines and transport helicopters were vital components of this nation's amphibious warfare capabilities and saw her involvement in the Cuban Missile Crisis of 1962 and the Dominican Republic intervention in 1965.

There would be two deployments to Vietnam wrapped around her brief involvement in the space program; namely, when her helicopters retrieved an unmanned AS-201 spacecraft in February of 1966 and her designation as primary recovery ship for the manned Gemini 8 mission a month later – a mission for which, it turned out, she was not needed. By December of 1969, Boxer was decommissioned and sold for scrapping in February 1971.

(Navy photo)

USS Wasp was again the recovery ship.

Enter "The Angry Alligator." More about that later but for now we'll just deal with lift-off and the earliest stages of this, the seventh manned, and the third rendezvous, mission of the Gemini program.

Agena target vehicles and their Atlas boosters were proving to be the biggest thorn in NASA's side and it continued with this mission. With astronauts Stafford and Cernan ready and waiting in their spacecraft on Launch Complex pad 19, their Agena took off and promptly blew up on its way to orbit. That might have dictated an inordinate delay in the next attempt but after an earlier mishap, NASA had directed the builder, General Dynamics/ Convair, to be ready to supply a back-up Atlas within two weeks of any more catastrophes. To complete a new package, if needed, NASA had an alternative to an Agena, the ATDA (Augmented Target Docking Adapter) in a nearby hangar, so that was quickly mated to the Atlas booster and the mission was back on track.

We mentioned earlier in this book that future astronauts would eventually pass their predecessors with regard to time spent sitting in a spacecraft waiting for a launch. Tom Stafford grudgingly grabbed those honors when, by the time what was now being called Gemini 9-A lifted off, he had been up and down the launch complex elevators and in and out of two spacecraft (Gemini 6 in December of 1965) no less than six times.

On June 1, 1966, at 8:39 a.m. the Atlas-ATDA combination rose from pad 14 and soon entered a near-perfect 185- mile orbit BUT this flight, too, came with its own immediate problem. Signals sent back to Mission Control from the target vehicle suggested that the protective launch shroud covering the actual docking port had not opened fully and had failed to jettison as it was designed to do. You may now start imagining what an angry alligator might look like. Astronauts Stafford and Cernan would get a first-hand look in just a few more days.

Gemini 9-A was relatively trouble-free when it lifted off from Launch Complex 19 on June 3 at 8:40 in the morning with Command

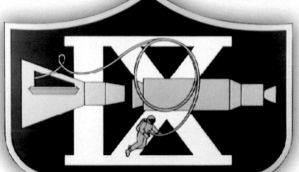

Pilot Stafford paying close attention to the spacecraft's newest gimmick, IVAR, which would guide the spacecraft in its rendezvous sequence. That procedure would begin a mere six minutes after launch when the ATDA was about 660 miles ahead. Stafford worked his thrusters to circularize their orbit and speed the spacecraft up to close with the target vehicle.

Soon enough they eased up on their target and confirmed Mission Control's suspicions. It was Stafford who put his description into the history books of space when he said, "It looks like an angry alligator out here rotating around." His immediate reaction was to use the Gemini spacecraft to bump into the ATDA and, hopefully, make those gaping shroud jaws behave. Back in Houston, flight director Gene Kranz told him not to. So, very early on, one of the mission's most important objectives was scrubbed. Investigation later underlined the old "too many cooks" theory when it was clear that a multiple of contractors had led to just one or two omissions that led straight to the ATDA malfunction.

Denied their chance to carry out a hard-docking, astronauts Stafford and Cernan turned to a series of rendezvous not unlike those that the Gemini 7 and 6 combo performed when they were not equipped to dock either. Gemini 9-A flew around the disabled ATDA and tested 3 different types of rendezvous including a simulation of what the upcoming Apollo spacecraft would have to accomplish enroute to a lunar landing. So, in other words, it was not a total loss but had to be totally frustrating for all involved.

And then there was mission goal number two. The second EVA (spacewalk) of the Gemini series and it was designed to be considerably more complex than the one carried out by Ed White during Gemini 4 (June, 1965). Wouldn't you think NASA deserved at least one trouble-free experiment after all the previous problems? Wrong.

The space walk assignment fell to Gene Cernan. It began only after those 3 rendezvous tests had been completed and Cernan was outside the spacecraft only a short time when he began to realize that everything, "every little movement," took a lot more exertion than it had in the simulators back in Houston. The Velcro added after earlier missions and the handholds that previous crews had suggested were just not enough. His umbilical cord, the only thing between Cernan and a grisly demise floating in space, came to be referred to by him as the "snake." Used at any distance at all, it became extremely difficult to handle.

And then…the dreaded foggy face-plate. All concerned later concluded the extra exertion required of Cernan caused his face-plate to fog up, making most of what he had hoped to accomplish – nearly impossible. He was unable to give the AMU a workout. That was the back-pack that was supposed to give astronauts more freedom and maneuverability on space walks. Very Buck Rogers…but "no sale" this time around.

Wouldn't you know the one experiment carried out successfully was the one all astronauts would come to dread. M-5 was the bio-assay of body fluids and it meant collecting and labeling, in laboratory fashion, all their body wastes during the mission. It was complex and it was messy and woe to the reporter who would try to press them for details on this particular phase of life in space.

They also were both able to take some spectacular photos in space but here again, both men would later report that even picture taking was more difficult than they had thought it would be. Pointing a camera and taking pictures while floating upwards in zero (or near-zero) gravity was no piece of cake.

On their 45th orbit they were packed and ready to head for home or at least the Atlantic. Since this was my first assignment as a "pool" correspondent (reporting for all the networks) aboard a carrier I had the privilege of standing in my broadcast position four levels above the flight deck of the USS Wasp and watching Gemini 9-A descend mere hundreds of yards from the ship under the beautiful, big orange-and-white canopy.

The helicopters, which flew from the deck of the Wasp, were from HS-11 and the frogmen came from the ever-present UDT-21, champions of Atlantic recoveries. The crew chose to stay inside the spacecraft and emerged only after it had been secured to its dolly on a starboard elevator on the Wasp's flight deck. Smiling and waving at all of us, you would have thought they didn't have a problem with any phase of their flight.

Even with their problems, which were thoroughly reported to the public via radio, TV and newspapers, the public was already starting to yawn simply because these flights were taking off and coming back with relative ease – at least to the casual observer. These phenomena would continue and it would be abetted by much less intense press coverage with each succeeding mission.

More on my personal involvement elsewhere in this book, but it continued to bother me on each and every recovery ship to which I was assigned. Here I was, a fortunate eyewitness to history, to good news as opposed to covering plane crashes (which I did anyway) and the public was steadily turning away. It was, and is, typical and probably unavoidable. It was the price of excellence or near-excellence. The public woke up, big time, only for the major space events such as the first manned mission to the Moon in 1969, the epic journey of the problem plagued flight of Apollo 13 in 1970, and the excruciating tragedies of the shuttle losses – Challenger in 1986 and Columbia in 2003.

Gemini 9-A is now one more attraction for you to see when you visit the Visitor's Complex at the Kennedy Space Center, Cape Canaveral, Florida.

Tom Stafford was an Air Force Colonel at the time of Gemini 9. Gene Cernan was a Navy Commander.

USS GUADALCANAL (LPH-7)

The USS Guadalcanal was the prime recovery carrier for two manned missions – Gemini 10 on July 21, 1966, (astronauts John Young and Michael Collins) and Apollo 9 on March 13, 1969, (astronauts Jim McDivitt, David Scott and Russell Schweickart).

The Guadalcanal was the first Iwo Jima class carrier to take part in manned spacecraft recovery. There would be five before the program ended and they would carry out nine recoveries. The basic LPH (Landing Platform Helicopter) was 592 feet long with an 84-foot beam and a draft of 27 feet. Initial displacement was 11,000 tons but full displacement was closer to 18,500 tons. She had a speed of 22 knots and accommodated 814 officers and crew…BUT…. could also carry 2,157 combat ready Marines and get them ashore quickly with her 25 helicopters. That was the basic LPH mission.

A BRIEF HISTORY OF THE USS GUADALCANAL (LPH-7)

Guadalcanal was the second ship of the U.S. Navy to be named after the first major offensive of World War II, which turned into a bitter six-month struggle and saw six major naval battles in waters surrounding that island.

The keel for the Guadalcanal was laid down on September 1, 1961, at the Philadelphia Naval Shipyard. She was launched on March 16, 1963, and commissioned on the 20th of July of that year.

Just months after her commissioning Guadalcanal set a new carrier record for helicopter landings in a single day – 1,134 – which shows just how quickly such a ship could land troops in a military operation.

In February of 1964 she was on station in the Panama Canal Zone for two months during unrest in the Panama Republic where the 8th Marine Expeditionary Unit embarked.

By the time she pulled into the shipyard of her birth, Philadelphia, in May of 1964 she had already logged over 10,000 helicopter landings.

On October 7 of that year Guadalcanal took part in Operation Steel Pike I off the coast of Spain. A NATO exercise, it was described as the largest amphibious training operation since World War II.

(Navy photo)

A number of other training exercises back in the Caribbean followed and in May of 1965, she was on hand to transport Army Air Units to the strike-torn Dominican Republic.

In July of 1966 she was selected as primary recovery ship for Gemini 10. Recovery in the Atlantic was accomplished on July 21.

The Guadalcanal would be part of Amphibious Ready Groups or Squadrons in the Caribbean and the Mediterranean from the mid-60's to the end of her naval service in 1994 with some very notable assignments, achievements and awards during those years.

On March 13, 1969, she picked up the three-man crew of Apollo 9, in the Atlantic as Gemini 10 had been nearly three years earlier. Future Apollo flights would splash down in the Pacific.

Into the seventies and eighties and Guadalcanal posted a presence in European waters from the Mediterranean and along the eastern coast of Europe up to Southhampton, England.

Two of her awards came as the result, in August of 1987, of a rapid deployment with the USS Okinawa in the Arabian Gulf as part of Operation "Earnest Will." The unit was credited with the first live moored mines swept (removed) by a U.S. Navy unit since the Korean War. That effort brought the Navy Unit Commendation and the Armed Forces Expeditionary Medal. October 7, 1993. U. S. Forces were ordered to deploy to the Indian Ocean off the coast of Somalia and

Guadalcanal was there as part of a three-ship amphibious ready group working in concert with a four-ship group bringing two Marine Expeditionary Units to that violent country, which inspired a best-selling book and only recently a film based on that book – *Black Hawk Down*. Anyone familiar with that story knows that a lot of Americans died there in clashes that saw a number of U. S. helicopters shot down by Somalian militia. The involvement would go on into early November.

The USS Guadalcanal, as part of that Joint Task Group, completed a six-month Mediterranean deployment early in February of 1994 and returned to its homeport in Norfolk, Virginia. She was decommissioned at the Philadelphia Naval Shipyard on August 3 of that year after 31 years of service and anchored for a while with the Atlantic Reserve Fleet there. Recent records indicate the Navy initially wanted to use Guadalcanal as a museum but the ship was last reported laid up in the James River Reserve Fleet at Fort Eustis, Virginia, and was scheduled to be sold for scrapping.

When those 31 years were over, in addition to the two commendations noted above, Guadalcanal had earned a long list of awards including the Navy Expeditionary Medal, the Humanitarian Service Medal, the Meritorious Unit Commendation and the Navy "E" Ribbon. Many of these awards were bestowed more than once.

THE FLIGHT OF GEMINI 10 WITH ASTRONAUTS JOHN YOUNG AND MICHAEL COLLINS FROM JULY 18 TO JULY 21, 1966

We could not know at the time of this mission that Command Pilot Young would not only go on to walk and drive around on the Moon (Apollo 16) but become the first shuttle pilot in 1981 and that his Gemini 10 colleague Michael Collins would be the guy who minded the store (Apollo 11 Command Module) while teammates Armstrong and Aldrin dropped down to have man's first look at…and walk on the Moon.

This would turn out to be a very exciting and successful mission and it would bring an amphibious assault ship (LPH), the USS Guadalcanal, into the picture as the first recovery carrier of that unique breed.

Once again NASA was determined to send an Agena target vehicle into space ahead of a Gemini spacecraft and to have a successful docking with it. Not only that but there was the other Agena still out there in space, the one which gave nightmares to the crew of Gemini 8. Both would be used successfully.

The launch of the Atlas-Agena target vehicle came at 3:39 p.m. on July 18 from launch pad 14. About one hundred minutes later, Gemini 10 roared from Launch Complex 19 and the "games had begun." The Agena assumed a near circular orbit while Gemini 10 slipped into an elliptical, or oval, orbit. Obviously the men in the spacecraft knew how to reconcile those differences because rendezvous was achieved at 5 hours and 23 minutes into the mission, and docking about a half hour later. That must have touched off some cigar lighting back in Mission Control with all previous attempts having failed.

Despite the joy of the first successful docking, so much fuel was used in achieving it that plans for repeating the procedure – docking and undocking a number of times – were scrubbed and an alternate flight plan had to be worked out. As it turned out, the Gemini 10 and the Agena (GATV) remained docked for 39 hours and the engines on the target vehicle were then used for six maneuvers to put the docked complex into position for a rendezvous with the Gemini 8 GATV, still out there but in a much higher orbit. This meeting was never planned as a docking but as a passive target, something to look at, check out and maneuver around. Besides there was a micrometeorite collection package attached to it and Mike Collins was able to go out on an umbilical EVA and bring it back. In the process, Collins used a nitrogen-fueled "space gun" to propel himself through space and became the first person in space to visit and make contact with a free-floating satellite which is what the Gemini 8 target vehicle had become. So you can see that Gemini 10 was fast becoming the darling of Mission Control and rolling up long sought after accomplishments in space.

We tend to lose sight of the fact that in every mission some sort of photography took place with varying degrees of success. Things kept getting better and when you are using cameras which the average person can only dream about, you can come up with some pretty fabulous photos. Hasselblad became the camera of choice but not the only one. And 70mm photography was put into play with often-awesome results. It doesn't make the camera buff feel any better to know that a few Hasselblad's were later left on the Moon to compensate for the weight of Moon rocks being brought back to Earth. A trade-off when total weight was a critical factor.

During EVA's the Gemini 10 crew conducted what was called Synoptic Terrain Photography and brought back invaluable pictures for geological, geographic and oceanographic study. Another photography experiment produced Ultraviolet Astronomical Photography, which opened up the heavens to stars previously possible to see and photograph only from unmanned rocket flights.

All in all a very nice trip. Retrofire was initiated at 70 hours, ten minutes after lift-off during the mission's 43rd revolution. Soon afterward, the spacecraft landed within sight (how about 3.4 miles?) of the USS Guadalcanal in the Atlantic and, this being my second "pool" reporting assignment in a row, I was watching the descent from my catbird spot on the carrier's 06 level and broadcasting it to the world. I was also watching helicopter squadron HS-3 at work, together with the frogmen of UDT-21 once again.

Where is Gemini 10 today? It's not here in this country but at the Norwegian Technical Museum in Oslo, Norway. Oddly enough, one of the spacecraft hatches can be seen at the Virginia Air and Space Center in Hampton, Virginia.

At the time of this mission, Michael Collins was an Air Force Colonel.

USS GUAM (LPH-9)

The USS Guam (LPH-9) was prime recovery ship for Gemini 11 on September 11, 1966. The astronauts were Pete Conrad and Dick Gordon.

Just like the Guadalcanal in the previous mission, Guam was an LPH so her statistics - measurements and performance – were much the same, give or take a ton.

A BRIEF HISTORY OF THE USS GUAM (LPH-9)

The USS Guam was the fourth ship of the Iwo Jima class and the third to bear the name Guam.

Her immediate predecessor was an 806-foot long cruiser, which became the flagship of Cruiser Task Force 95, earned two battle stars for World War II service, and was decommissioned on February 17, 1947.

She was built at the Philadelphia Naval Shipyard where her keel was laid on November 15, 1962. She was launched on August 22, 1964, and completed March 31, 1965.

She became the flagship for Amphibious Squadron 12 and devoted most her time plying the Caribbean Sea and the waters of Central America.

For a ship built in the post-war period, Guam managed to put together a commendable service record and, on more than one occasion, found herself "in harm's way."

Nonetheless, very early in her naval career, Guam was designated as prime recovery ship for a manned space flight – Gemini 11, which splashed down in the Atlantic on September 18, 1966. More on that mission – coming up.

Ten years later came the distinction of serving as a floating "Federal District Court" when, during our bicentennial year, 200 new U.S. citizens were naturalized on Guam's hangar deck.

In May of 1982 while deployed to the Mediterranean, Guam was sent to the coast of Lebanon for possible evacuation operations as war raged between the Israelis and opposing Palestinian and Syrian forces. The ship took part in the evacuation of over 600 Lebanese, Americans and third country nationals from Juniyah, Lebanon, a city north of Beirut. For her efforts, Guam received the Navy Unit Commendation and the Humanitarian Service Medal.

It was a busy year for Guam because in August she performed exactly as designed when her helicopters landed combat-ready Marines in Beirut as part of the multi-national peace keeping force that included French and Italian troops. The ship then participated in evacuation of PLO guerrillas

(Navy photo)

from Beirut, left briefly only to return in September to re-deploy Marines.

In 1983 Guam provided logistic support during the invasion of Grenada and took part in the rescue there of 200 American citizens.

A grim assignment came in early 1986 when Guam took part with many other ships in retrieving parts of the ill-fated shuttle Challenger, which claimed the lives of all seven of its crewmembers.

In February 1991 Guam supported amphibious operations during Operation Desert Storm in Iraq…. a conflict, which began with ground forces on February 23 and ended four days later. That same year Guam helped evacuate American personnel from Somalia in Africa.

Guam was decommissioned on August 25, 1998, and laid up in the James River Fleet at Fort Eustis, Virginia. While many ships of the line end up in the nation's scrap yards, Guam was destined for a different ending. Some might argue that while being sunk during target practice and weapons testing was, somehow, her last service to her nation, it is still sad to write that she was sent to the bottom about 400 miles off the coast of North Carolina on October 16, 2001.

In addition to the awards given to the USS Guam for her work during the crisis in Lebanon, she also earned the Armed Forces Expeditionary Medal, the Navy Expeditionary Medal and a second Humanitarian Service Medal.

THE FLIGHT OF GEMINI 11 WITH ASTRONAUTS PETE CONRAD AND DICK GORDON
FROM SEPTEMBER 12 TO SEPTEMBER 15, 1966

NASA had gotten its longest flights from the missions of Gemini 4, 5 and 7 and they had proven man's ability to function in space. Now it appeared that other issues were moving up in priority and chief among them was rendezvous and docking – so vital to any Moon mission. There had been some well-reported bumps in that road but things were starting to fall into place and the Gemini series would produce an excellent wind-up and serve as the table setter for Apollo.

Gemini 11 also saw the appearance of an amphibious assault ship (helicopters only), the USS Guam, for the second straight mission. It would not be the last.

Astronauts Conrad and Gordon rocketed into space at 9:42 on the morning of September 12, about an hour and a half after their Agena target vehicle (GATV-5006) had gone into orbit. As usual, the Agena left from Launch Complex 14 and Gemini 11 from pad 19.

Proving the crew and the equipment were up to their assignment, rendezvous was achieved just an hour and 25 minutes into the mission and docking just 9 minutes later. Clearly, the lessons learned from the prior flights were

paying off. The crew then began what NASA had wanted so badly for a long time – docking and undocking practice. Each man would dock, undock and dock again in both daylight and darkness and Conrad reported it was easier than their ground simulations.

After the last docking, the crew fired the Agena main engine for a test prior to their high-altitude run. It was Charles "Pete" Conrad who had lobbied hard for it from the time he was in training the year before.

He had even backed a plan that would have seen a Gemini go to the Moon and back (but, of course, no landing). NASA's top leadership shot that one down but Conrad kept up his crusade for deep space and it became part of Gemini 11's flight plan.

Before the ride into deeper space there was an EVA by Dick Gordon and, as had been the case several times before, Gordon found the effort required outside the spacecraft was very, very difficult – exhausting - and his co-crew member could see it as well. How tired was he? Well the EVA that had been scheduled to last 107 minutes had to be terminated at 33 minutes. Even the

USS Guam on her way to rendezvous with Gemini 11. (Navy photo)

primarily from a stand-up EVA. All he was doing was standing up on his seat with the hatch open and the camera clicking away.

They had problems with dirty windshields, which proved difficult to clean. Not like your basic gas station operation…. not in space anyway. They also had problems familiar to previous missions with the umbilical between them and the Agena. It seemed to have a life of its own and was constantly loosening, tightening up and loosening again.

End of mission was approaching so it was time to separate from the target vehicle, after which Conrad used the thrusters to slow the spacecraft down, taking it to a lower orbit while they continued to keep an eye on Agena. When it was finally time to put some space between them and the target vehicle, the crew did so with great reluctance. Conrad would say later it was the best friend they ever had and Gordon would add that they were "sorry to see that Agena go. It was very kind to us."

Just about the last significant event left to the Gemini 11 crew was an automatic re-entry; and while both men watched, the computer did its job after a few almost human hesitations. The system was able to recover from whatever was bothering it, use a minimum amount of control fuel, which was the whole idea, and plop the spacecraft down within 2.7 miles of the recovery carrier USS Guam where the same helicopter squadron (HS-3) and frogman unit (UDT-21) we had seen just a few weeks earlier on Gemini 10, did the job. I was up there in my favorite broadcast position on the ship's 06 level. I would not sail aboard a recovery ship again for nearly three years; but when I did, it was the one mission we all wanted – Apollo 11.

Now then — where to see the Gemini 11 spacecraft. It's at the California Museum of Science and Industry in Los Angeles.

Dick Gordon was a Navy Captain.

handholds and Velcro fasteners added after previous missions were not good enough.

A flight controller on a tracking ship sent them up the numbers for their next big event…. the high ride, after which the astronauts tidied up their home, had dinner, took some pictures and then hunkered down for some badly needed rest…especially for Dick Gordon.

So it was that 40 and a half hours into their mission and on the 26th orbit of Earth, Pete Conrad pushed a button that started the main engine of the attached Agena target vehicle and it was quite a blast. It fired for nearly 30 seconds and greatly increased their speed, so much so that both men were thrown forward in their seat harnesses even as they watched the Earth get smaller and smaller. Conrad was sending back comments on the view like an awe struck tourist but not so much that he and Gordon forgot about photography. They took over 300 pictures during the "high ride."

For the remainder of their mission Conrad and Gordon gave their cameras a continued workout and Gordon did his picture taking

USS Wasp was again the recovery ship for the fifth and last time.

This one would prove to be a most happy ending to the ten mission Gemini series with Buzz Aldrin, often described as one of the most engineering-savvy members of the astronauts' corps, showing how an EVA could, and should, be done. Aldrin could point to doctoral studies at MIT. He came to be called "Doctor Rendezvous" by McDonnell-Douglas (builders of the Gemini) and MSC engineers. For this last Gemini mission, they had the right guy in the right place.

Main objectives continued to be gaining experience in rendezvous and docking. NASA never felt that the program could get too much of that. They wanted another high-apogee deeper space trip but that was soon scrubbed because of a malfunction noted in the Agena target vehicle's primary propulsion system during its entry into orbit.

While Buzz Aldrin could certainly be classified as a "brainy" sort of guy, he and fellow crew member Jim Lovell were not humorless. Indeed, just minutes before the Agena target vehicle roared off Pad 14, Aldrin and Lovell had made their appearances on the ramp leading up to Pad 19 and Aldrin had "THE" on his back while Lovell had "END" on the back of his space suit. Considering the risky business they were all in, this could have been considered a bad joke but it was accepted not just because everyone was so confident, but that Gemini 12 was clearly "the end" of the Gemini program. This fact of space life was punctuated five days earlier by the launch of the unmanned Lunar Orbiter II, which headed for the Moon and took many photos of potential Apollo landing areas.

Lift-off for Gemini 12 occurred at 3:46 p.m., roughly 90 minutes after Agena headed out. After a communications dropout and some anxious moments of unwanted silence, the ground tracking station at Tenerife broke through with data the crew would need for docking with the Agena – coming right up.

That was followed by a radar failure and this is where "Doctor" Aldrin began to show his stuff. You know what a sextant is? It's an instrument long used by navigators to use stars, the sun and such to determine a ship's position. The dictionary tells me it looks like "one-sixth of a circle." Got that? Well, Aldrin had already realized that radar wasn't going to get them up to the Agena and he had NASA's modern version of the sextant out and was using it, effectively, to do with his hands, mind and eyes what modern technology had decided not to do. It worked and they docked with Agena at 4 hours, 13 minutes into the flight. A beautiful example of "seat of the pants" navigation, but since it was now the fourth flight to accomplish a docking, comment from the ground was a matter-of-fact "Roger."

Then, and for only the second time in the Gemini series, they were able to practice docking and undocking. A few blips and glitches but everything eventually worked out. Next…firing up the Agena's engine to head deeper into space but, as we noted here earlier, an Agena malfunction had already been noted, checked out and now convinced flight directors back in Houston to not even try the maneuver.

Consequently, the prospects of photographing a solar eclipse, which had been sidelined after a two-day launch delay, now crept back into consideration and after much give-and-take… it took… it stayed in the mission. It was some 16 hours into the mission when they saw it "right on the money" and took some incredible photographs.

There was additional photography as Aldrin did a stand-up EVA and joined previous Gemini flyers in his honest and enthusiastic amazement at what he was privileged to observe. There was ultra-violent astronomical picture taking and even retrieval of a micrometeorite collection package, and it was all part of a highly successful 2 hour and 20 minute exercise.

The next day, or as close as you can come to distinguishing day from night when you are orbiting the Earth every 90 minutes, came the main mission event. Could an astronaut really perform useful tasks at the end of one of those frisky, unpredictable umbilical cords?

Remember, that lifeline had already worn out some pretty good men. Aldrin, doubtlessly benefiting from the previous experience and recommendations of predecessors, nonetheless proceeded to deliver a text-book performance as he used a handrail to maintain his position outside, a waist tether to give him two free hands and the new so-called golden slippers that allowed him to slide his booted feet easily into and out of these restraints. He found he was able to not only relax, but also lean as much as 45 degrees to either side or 90 degrees backward.

What followed shortly was working with what NASA had named a "busy box" not unlike what a youngster might be exposed to in teaching him or her the fundamentals of picking things up, turning those things or tightening something. Aldrin had to do some exercises that would have appeared childlike here on solid ground but which, in space, were proving to be major challenges. At one point, Aldrin drifted over to the target vehicle as Lovell watched him pull some wiring apart and put it back together. Fast-forward to the International Space Station where such dexterity is an absolute must for completing that "city in space" and you can easily see the need to know. The sense of humor arose again as Aldrin headed back to the Gemini and stopped to wipe the windshield. Lovell asked him about checking the oil and the air in the tires but was told they were O.K., at which point Aldrin climbed back aboard.

The third EVA, and the second of the stand-on-your-seat variety, came on the fourth day of the mission and ran about an hour. They tossed out a lot of unneeded "stuff" but lest you think they were cluttering up outer space, be advised that in this case and in all previous "dumping" exercises, that "stuff" was quick to suffer orbital decay, fall back into our atmosphere, and burn up.

There were a few fuel cell problems, too much water in the service module's tanks, but like most anomalies encountered on missions, they were conquered and recorded for others to learn.

Twelve out of fifteen experiments yielding most of what was expected of them isn't bad and, besides, a lot of ingenuity had been necessary to overcome the various problems. Aldrin was "Mr. Cool," and long before he became the second person to walk on the Moon, his Gemini 12 performance had been described as flawless, chock full of ease and style.

During the 59th revolution Gemini 12 began its controlled automatic re-entry. It went well except for a messy situation that happened when a storage pouch came loose from its Velcro straps and everything from books, pamphlets and bits of hardware broke loose at a time when they were trying to focus all their attentions on getting back down to Earth. They dealt with it and landed even closer to the recovery ship, the carrier Wasp, than Gemini 11 had to the carrier Guam – just 2.6 miles. The frog teams (UDT-21) dropped into the Atlantic from their choppers (HS-11) and they had the astronauts on Wasp's deck in less than a half hour after what had been a hard splashdown.

It had been less than 5 years since Alan Shepard's 15-minute sub-orbital flight and a lot of frontiers had been reached and crossed. Rendezvous and docking, space walks, deep-space missions, extended time in weightless space, fabulous picture taking, space-age foods, and computer-controlled re-entries. They'd been there and done all that. Bring on the Apollo program.

As far as getting a look at Gemini 12, you will find this spacecraft at one of NASA's most storied sites for research and development, the Goddard Space Flight Center in Greenbelt, Maryland.

At the time of this mission, Buzz Aldrin was an Air Force Colonel.

GEMINI SUMMARY

MISSION	LAUNCH DATE	LAUNCH TIME (Z)	HIGHEST APOGEE (MILES)	LOWEST PERIGEE (MILES)	NUMBER OF REVOLUTIONS	LANDING DATE	LANDING TIME	LANDING COORDINATES	MISS DISTANCE (MILES)	ASTRONAUTS	MISSION DURATION (HOURS)	PRIME RECOVERY SHIP	TIME TO RECOVER ASTRONAUTS	TIME TO RECOVER SPACECRAFT	DOD RESOURCES MEN	DOD RESOURCES AIRCRAFT	DOD RESOURCES SHIPS
GT-1	8 Apr 64	1600	173	87	64	-	-	-	-	Unmanned	3 orbits	-	-	-	5,176	11	3
GT-2	19 Jan 65	1404	92	-	-	19 Jan 65	1422	16°36'N 49°46'W	34	Unmanned	:18	CVS Lake Champlain	-	1:30	6,562	67	16
GT-3	23 Mar 65	1424	121	88	3	23 Mar 65	1917	22°26'N 70°51'W	60	Grissom Young	4:52	CVS Intrepid	1:12	2:47	10,185	126	27
GT-4	3 Jun 65	1516	160	86	62	7 Jun 65	1712	24°44'N 74°11'W	44	McDivitt White	97:56	CVS Wasp	:57	2:16	10,249	134	26
GT-5	21 Aug 65	1400	189	87	120	29 Aug 65	1255	29°44'N 69°45"W	92	Cooper Conrad	190:55	CVS Lake Champlain	1:31	3:55	10,265	114	19
GT-6A	15 Dec 65	1337	168	87	16	16 Dec 65	1528	23°35'N 67°50'W	7	Schirra Stafford	25:51	CVS Wasp	1:04	1:04	10,125	125	16
GTA-7	4 Dec 65	1930	177	87	206	18 Dec 65	1405	25°25'N 70°07'N	6.4	Borman Lovell	330:35	CVS Wasp	:32	1:03	10,125	125	16
GTA-8	16 Mar 66	1641	161	86	7	17 Mar 66	0322	25°14'N 136°44'W	1.1	Armstrong Scott	10:41	DD Leonard F. Mason	3:06	3:15	9,655	96	16
GTA-9	3 Jun 66	1340	168	86	455	6 Jun 66	1401	27°52'N 75°00'W	0.4	Stafford Cernan	72:21	CVS Wasp	:52	:52	11,301	92	15
GTA-10	18 Jul 66	2220	412	86	43	21 Jul 66	2107	26°45'N 71°57'W	3.4	Young Collins	70:47	LPH Guadalcanal	:27	:52	9,067	78	13
GTA-11	12 Sep 66	1442	741	87	44	15 Sep 66	1359	24°15'N 70°00'W	2.7	Conrad Gordon	71:17	LPH Guam	:24	:59	9,054	73	13
GTA-12	11 Nov 66	2047	163	87	59	15 Nov 66	1922	24°34'N 69°55'W	2.6	Lovell Aldrin	94:34	LPH Guam	:30	1:07	9,775	65	12

After more than thirty years there is no way I could recall where this handy little Gemini mission summary chart came from but it has been "among my souvenirs" all this time nonetheless. Interesting facts and figures. I wish I had a similar chart for both Mercury and Apollo. (Note: chart re-created for clarity)

USS ESSEX (CVS-9)

The USS Essex was the primary recovery ship for the Apollo 7 Earth orbital mission of Navy Captain Wally Schirra, Air Force Major Donn Eisele and Marine Major Walt Cunningham on October 22, 1968.

Here she is. The ship that gave a name to a whole class of carriers, a much-decorated "short hull" carrier that capped a stellar naval career with recovery of the first manned Apollo flight. Essex measured 872 feet in length, beam of 93 feet, a top speed of 33 knots and crew complement of 3,448.

A BRIEF HISTORY OF THE USS ESSEX (CV-9)

The Essex, which picked up astronauts Schirra, Eisele and Cunningham, was only the fourth ship to bear that name.

The Essex class carrier was the most dominant naval force in World War II as it became rapidly apparent that our aging battleships would not get the job done, particularly in the Pacific.

Essex (CV-9) was built at the Newport News Shipbuilding and Drydock Company, launched on July 31, 1942, and commissioned on New Year's Eve of that year.

Having been built relatively early in the war Essex would go on to become one of our most honored ships, receiving the Presidential Unit Citation and 13 battle stars for World War II, plus 4 more battle stars and the Navy Unit Commendation for Korean war service.

War began for Essex when she joined Task Force 16 in carrier operations against Marcus Island in August of 1943, struck Wake Island in October and then became part of the landing at Tarawa in the Gilbert Islands in late November. Her second amphibious assault came in the company of Task Group 58.2 against the Marshall Islands in late January and early February of 1944.

Only a few weeks later Essex joined two other Task Groups to launch a massive attack on the island of Truk during which eight Japanese ships were sent to the bottom. Then it was off to support attacks on Saipan, Tinian and Guam during the third week in February.

USS Essex – before angled deck. (Navy photo)

Following that action it was time for her single wartime overhaul and she sailed back to San Francisco. By mid-May Essex was back in action, joining the USS Wasp and the USS San Jacinto in a naval group attacking Marcus and Wake Island again. Other action saw her in the Marianas, the Palau Islands and Mindanao in the Philippines. Enemy shipping was the main target there.

For the remainder of 1944 she took part in strikes on Okinawa and Formosa and in the epic Battle for Leyte Gulf. On November 25, Essex absorbed her first kamikaze hit with the suicide pilot bringing his aircraft right down into some of the ship's planes being gassed up for takeoff. Damage was extensive, 15 men were killed and 44 wounded.

Despite that terrible blow, repairs that followed were rapid, and by mid-December Essex was back in the thick of things, dodging several typhoons over the next six weeks while participating in strikes against Formosa, Sakishima, Okinawa and Luzon.

For the remainder of the war she operated with famed Task Force 58, conducting raids on Tokyo to neutralize Japan's airpower before landings on Iwo Jima and Okinawa. Her aircraft were pounding targets in the Japanese home islands right down to the last days of the war in the Pacific.

In September 1945, she was ordered to Bremerton, Washington, for inactivation and on January 9, 1947, she was placed out of commission and in reserve.

No rest for the battle weary. In January of 1951 Essex was modernized with a new flight deck and a streamlined island superstructure and was soon to begin the first of three tours in Far Eastern waters during the Korean War. That new flight deck was needed because Essex was to become the first carrier to launch the new F2H Banshee twin-jet fighters on combat missions. One of those planes, damaged in combat, crash-landed on her deck and it cost the lives of 7 men.

Training exercises with the 7th Fleet kept Essex busy from November 1954, to June of the following year. A month later she would be at the Puget Sound Naval Shipyard for alterations, which included the angled flight deck most of her wartime sister ships were in the process of getting. This modernization prepared her for a new role as an anti-submarine warfare support carrier, her primary mission for the rest of her active naval life well into the 1960's.

An entirely different assignment came up in the fall of 1968, as it did for many Essex class carriers, the role of prime recovery ship, and this time for the Apollo 7 mission.

You could say that deployment was the crown on a long, legendary career because by June of 1969 Essex was decommissioned, stricken from the Navy List on June 1, 1973, and sold for scrapping in June of 1975.

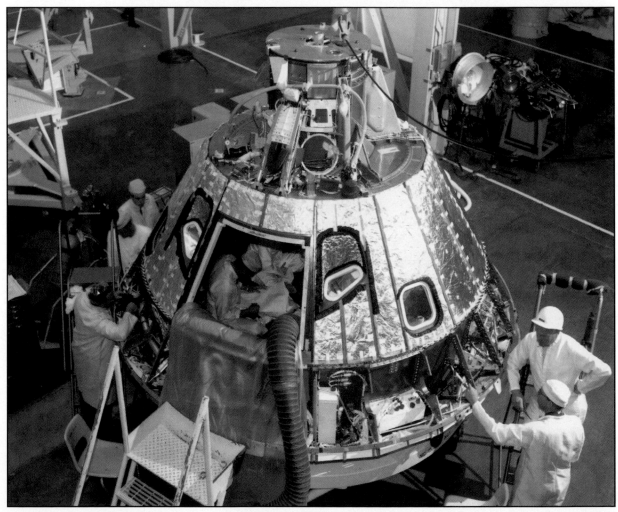

Engineers at North American Aviation in Downey, California, putting the finishing touches on an Apollo Command Module sometime in the mid-1960s. (Courtesy of NASA)

Time for Apollo. It began with tragedy and ended in triumph. The horrible launch-pad fire, which took the lives of astronauts Grissom, White and Chaffee on January 27, 1967, stopped the space program in its tracks. The crew perished in a pure oxygen environment, which was immediately tossed aside and flight would not resume until that danger had been totally eliminated.

It would be 20 months later before NASA was ready to try again. The astronauts would breathe air that was now 65% oxygen and 35% nitrogen. Certain flammable interior segments had also been modified or replaced.

Back in the spacecraft for the first time in nearly three years (Gemini 6-A, Dec, 1965) was the always up-front Wally Schirra along with space rookies Donn Eisele and Walt Cunningham.

An Apollo three-man spacecraft was a quantum leap in size and weight and mandated a much more powerful launch vehicle. Enter the Saturn series, designed and built by former German rocket scientist Werner Von Braun and his team of former associates operating out of the Marshall Space Flight Center in Huntsville, Alabama. This time it was Saturn 1B, good enough for the initial Apollo orbital flights but not good enough for the eventual lunar missions when the incredibly powerful (and still champion) Saturn V would make its debut. More on Saturn V later, but while 1B was 140 feet tall; the complete Saturn V complex would be nearly 365 feet in length.

This would also see launches taking place at Launch Complex 34, the same site where the Apollo 1 astronauts had died in January of 1967. The Gemini launch sites had been taken down soon after that program ended in November of 1966.

The crew would be taking along, for the first time, the Apollo Command Service Module which elongated the usual spacecraft profile but which contained vital equipment such as oxygen, water, consumables and hydrogen. The Lunar Module, which would later take our astronauts down to the Moon's surface, was not needed for this essentially practice flight.

A perfect lift-off at 11:02 a.m. Five minutes later they were comfortably in an elliptical orbit from about 140 to 183 miles above the Earth.

Even though this mission was to last nearly 11 days, it was not chock full of experiments like so many missions that preceded it. Its main purpose, apart from the always important study of man's ability to function in a weightless environment for varying lengths of time, was to test and retest and test again the Service Module's propulsion system which absolutely had to work to get the CSM (the combination of Command and Service Module) in and out of lunar orbit. Out of eight attempts there were eight nearly perfect firings. The only surprise came on the first firing when, to the crew's surprise, the propulsion engine kicked like a mule. Wally Schirra, never at a loss for words, yelled "Yaba-daba-doo" with thanks to Fred Flintstone.

After they had done some station keeping with their second stage rocket, the S-IVB, for about one and a half orbits, Walt Cunningham, who played the role of Lunar Module Pilot even though there was no Lunar Module along on this mission, told Mission Control that the adapter panels designed to mate the Lunar and Command Modules, had not fully deployed. It just so happened that Tom Stafford was on the back-up crew in Mission Control and it instantly reminded him of the "angry alligator" balky target vehicle that he had to cope with on Gemini 9 back in June of 1966. Not a major problem since those panels could, and would be, explosively jettisoned on future flights.

Once again the astronauts had real, unpleasant problems with the waste disposal system on the spacecraft. Problems which, though duly noted here, will not be dealt with in detail. Aren't you glad?

They had some fuel cell and battery problems but never to such an extent that the mission was endangered.

Since Wally Schirra had broken out with a bad cold some 15 hours into the mission and his crewmates could not help but catch it too, there was a whole lot of concern about wearing their space suit helmets upon reentry. That would have prevented them from blowing their noses and in weightless space the problem is simply compounded. Down in mission control, Deke Slayton tried to talk them into keeping the helmets on but Schirra was adamant and his word carried the day.

Despite their colds and some grumpiness probably brought on by those cold germs, the three men became Earth orbit's first TV stars with no less than seven live transmissions back to the folks on Earth. Those "shows" attempted to show viewers a different phase of, or part of, the spacecraft each time around. It was not crystal clear television but it was a start and, like the food, it was destined to improve with each flight.

As had become the norm, there was plenty of still photography going on. They also had both hot and cold drinking water on board for the first time thanks to the fuel cells in the service module.

NASA had to be very pleased. No really serious problems and the mission was capped with a soft splashdown in the Atlantic southeast of Bermuda at about 12 minutes past 7 a.m. on the 22d of October. They came in at about 2 miles from the so-called target point and only a few miles further from the primary recovery ship, the carrier USS Essex which was about to carry out its only recovery assignment. Captain J. A. Harkins commanded Essex. HS-5 was the helicopter squadron on board and its Recovery helicopter pilot was Commander E. A. Skube. UDT-21 continued to be the workhorse among the swimmers, having done all the Gemini Atlantic recoveries.

Frogmen, dropping from HS-5's helicopters, stabilized Apollo 7 and had the crew on the deck of Essex less than an hour after splashdown and the Command Module on board about an hour later.

We have not followed every spacecraft beyond the usual offload at Boston, Norfolk, or later on at Pearl Harbor, but in this case we'll simply point out that Apollo 7 was offloaded at Norfolk on October 24, that evaluation and deactivation procedures were completed by October 27. Then the spacecraft was shipped to Downey, California, for post-flight analysis by its builders – North American Rockwell Space Division.

If you were keeping score, Apollo 7 was out there for 10 days, 20 hours, 9 minutes and 3 seconds. It's in the books. Plus, they orbited this old Earth 163 times.

And where is Apollo 7 today? At the National Museum of Science and Technology in Ottawa, Canada.

Donn Eisele was an Air Force Colonel and Walt Cunningham was a Marine Major at the time of this mission.

USS YORKTOWN (CVS-10)

The USS Yorktown was the primary recovery ship for the Apollo 8 mission of astronauts Frank Borman, James A. Lovell Jr., and William Anders from December 21 - 27, 1968. That flight will forever be remembered as the Christmas mission during which the astronauts circled the Moon and sent back moving holiday messages and observations.

Yorktown proved that there are exceptions to every rule. Yes, she was built as, and remained, Essex class throughout her naval career which was nothing short of spectacular. Unlike other "short hull" Essex class carriers such as Intrepid which remained straight deck ships – Yorktown was converted to an angled deck in the mid-1950's but that length (872 feet) and other Essex characteristics did not change.

Yorktown at Patriot's Point Museum, Charleston, S.C.

A BRIEF HISTORY OF THE USS YORKTOWN (CVS –10) THE FIGHTING LADY

Yorktown, like the Essex, was the fourth ship to bear the name and her keel was laid down just 6 days before the attack on Pearl Harbor – December 1, 1941. It illustrates the time required to build a carrier in that Yorktown was not launched until January of 1943 nor commissioned until April 15 of that year. At the time the keel was put down, the name of the ship was to have been Bon Homme Richard, but on September 26, 1942, the Navy decided she would hit the water as Yorktown.

Her statistics, displacement, length, speed and crew complement were identical to the Essex.

Being born so early in the war, Yorktown was destined to play a major role in the war in the Pacific. By the time World War II ended, she had received the Presidential Unit Citation and 11 battle stars. She missed the Korean War due to a major overhaul but was awarded 5 battle stars for service during the Vietnam War.

Carriers such as Yorktown and Essex along with Enterprise, Wasp and Hornet would share many experiences as part of the famed naval Task Forces, which figured so prominently in the Pacific campaign. The reader may well imagine several of these mighty ships sailing within sight of each other as their aircraft carried out raids on targets such as Marcus and Wake Island in the fall of 1943.

Her first major assault operation came with Task Force 50 of the so-called Fast Carrier Forces in the Gilbert Islands on November 10. Some very unfamiliar island names cropped up in her battle records as her planes flew against targets on Jaluit and Mili, Abemama and Makin Islands. The name Tarawa is much more recognizable and was another target.

Yorktown was sent back to Pearl Harbor early in December to render badly needed air training for new pilots coming out. By January 16, 1944, she was heavily involved in an amphibious assault operation called Operation Flintlock in the Marshall Island with the Lexington and other ships in Task Group 58.l. One of the islands hit hard and often was Kwajalein.

Over the next four months Yorktown took part in a series of raids ranging from the Marianas in the north to New Guinea in the south. One of those attacks had the main Japanese anchorage at Truk as a key target, and it wouldn't be the last.

In May Yorktown would find herself back at Pearl Harbor for more pilot training operations. By the 29th of the month she was headed back into the western Pacific for the upcoming assault on the Marianas. Her planes took off early in June to soften up targets for the invasion of Saipan and that meant hitting Japanese airfields on Guam.

Reunited with Task Force 58, Yorktown then became a key player in the Battle of the Philippine Sea, claiming 37 enemy planes downed on the first day of that famed battle. A series of attacks on islands such as Iwo Jima and Chichi Jima followed but by the 31st of July, Yorktown was headed back to Pearl Harbor for a two-month overhaul.

In November, and now attached to Task Force 38.1, she began launching air strikes in support of the Leyte invasion in the Philippines. As that operation was ending and the Task Force was withdrawing, it ran into a vicious typhoon, which claimed three of our destroyers. Yorktown then became part of the rescue operations for survivors of those sinkings.

What followed in the next few months were attacks on Japan's inner defenses and hunts for elements of the Japanese fleet in those waters. Raids were directed at targets on Formosa, the Japanese mainland and Canton and Hong Kong in China.

Throughout these operations Yorktown and other elements of the various task forces were in and out of the then virtually unknown anchorage at Ulithi. As noted elsewhere in this book, it was purposely so, and the American public hardly heard the name since Ulithi represented the largest and best anchorage for repairs and replenishment for nearly a thousand miles in any direction. It wasn't until after the war that the Navy began to talk about the island and the role it the played in the Pacific.

(Navy photo)

As 1945 began, the island-hopping campaign was closing in on the Japanese home islands and Tokyo was on the target list more and more, even as attacks continued in support of upcoming landings on Iwo Jima and Okinawa.

It was early in April when intelligence reported a Japanese task force steaming south for one last desperate offensive and the super battleship Yamato was a part of that force. Yorktown and other carriers quickly launched attacks against that valued target and sent the big battlewagon to the bottom.

For several more months Yorktown sortied with several Task Groups and their assaults on Japan now included the northernmost island of Hokkaido. During these operations the only two atomic bombs ever used in wartime leveled the cities of Hiroshima and Nagasaki and on August 15 Japan decided to accept Washington's demands for unconditional surrender.

Not surprisingly then, the Yorktown and other carriers of Task Force 58 steamed around in waters east of Japan waiting for instructions on what they should do next. Warriors suddenly without a war to fight. Soon they were involved in air-dropping supplies to Allied prisoners of war still living in their prison camps and bringing personnel back to the states.

There were several trips from Guam and the Philippines back to San Francisco before Yorktown arrived in Bremerton, Washington, in late January of 1946. By June she had been placed in commission, on reserve. On January 9, 1947, she was taken out of commission and berthed with the Pacific Reserve Fleet at Bremerton.

Yorktown remained in reserve for almost five years but renovation and modernization was initiated in the summer of 1952. She was back in full commission by February 20, 1953. By the time she transited to Yokosuka, Japan, in September, the Korean War armistice was already two months old so Yorktown was not to have a role in that conflict at all. Instead she headed for home, some brief repairs and a role in the film "Jet Carrier." Then it was back to the Far East and maneuvers with the 7th Fleet.

In March of 1955, back at the Puget Sound Naval Shipyard, Yorktown finally got the angled deck that most of her sister ships were undergoing to increase their jet aircraft launching capability. She would put in a lot of time with the 7th Fleet over the next several years but by September of 1957 her home port was changed from Alameda to Long Beach and she was re-classified an anti-submarine warfare (ASW) carrier and received her final designation – CVS-10.

In January of 1960 she earned her initial recognition for duty in Vietnamese waters with the awarding of the Armed Forces Expeditionary Medal.

The next few years were taken up with several anti-sub warfare exercises with members of SEATO…the Southeast Asia Treaty Organization. In 1964 and 1965 Yorktown saw her first real involvement in the Vietnamese war, providing anti-submarine services for the fast carriers conducting air strikes on Vietnam. She would conclude her service in that theater with five more battle stars to her credit.

On February 24, 1967, Yorktown entered the Naval shipyard at Long Beach, California, for a seven-month overhaul and when she returned to the Pacific she provided search and rescue support for the contingency force assembled in the wake of North Korea's capture of the Pueblo (AGER-2) early in 1968. That mission was followed by three more tours of duty with Task Force 77 on Yankee Station and further involvement in the Vietnam fighting, primarily ASW support for the fast carriers launching the air strikes there.

On July 5 of that year she entered the Long Beach Naval Shipyard for three months of repairs and then returned to her normal operations but with a few unique assignments coming right up. First she served as a platform for the filming of the Pearl Harbor re-creation "Tora! Tora! Tora!" in late November and early December. A short time later she was named the prime recovery ship for the Christmas flight of Apollo 8, which circled the Moon and splashed down in the Pacific on December 27.

Together, with her continuing ASW assignments, Yorktown then became a frequently seen visitor at various European ports in Germany, England, France, the Netherlands and Denmark. It was like a farewell tour because deactivation and decommissioning was right around the corner. She went into the Philadelphia Reserve Fleet in June of 1970 and remained there for three years before the Navy Department approved her donation to the Patriot's Point Development Authority in Charleston, South Carolina, as a floating museum, where she remains to this day.

The author easily recalls seeing an excellent documentary entitled "The Fighting Lady" at a neighborhood movie theater (Irvington, N.J.) in 1944 when he was an impressionable 11 year old. That film is now showing several times a day at the Yorktown Museum.

The Christmas mission. The first manned mission to the Moon and back. No landing on the Moon itself, of course, but even if Grumman Aircraft on Long Island had been able to turn a LM (lunar module) over to NASA in time, the space agency was not ready to go for a landing on the Moon just yet. They might have taken the LM along for a whole lot of other reasons but since delivery would not be made for a number of months, NASA turned Apollo 8 into the exciting and memorable round-trip to the Moon that it became. The fact that it unfolded over the Christmas holidays only made it more unforgettable.

As soon as Apollo 8 became a trip to the Moon, and not just an Earth-orbital check-out type of mission, NASA had to roll out the mighty Saturn V since Saturn 1B, which had sent Apollo 7 into Earth orbit, had nowhere near the power necessary to escape from Earth's gravity. Saturn V was then, and remains, the most powerful rocket ever built although it is believed the Soviet Union had one that was even more powerful but which exploded on a launch pad at Baikonur, the Soviet space launch complex.

Even with secrecy-shrouded disaster, rumors were making the rounds that the Soviets were getting ready to send a manned spacecraft around the Moon so there was no shortage of urgency in Houston in those days. In addition, Moscow had sent a couple of unmanned spacecraft to the Moon in recent months but not into lunar orbit. In other words, NASA folks were looking over their shoulders a lot. With the arrival of Saturn V on the scene, launch activities moved to the new complex at Launch Pad 39A and lift-off occurred at 7:51 a.m. December 21, 1968. For nickname fans, bad news, this break-through mission had no nickname for the spacecraft.

To escape the gravity of this Earth, how about a top speed of 24,593 mph? That would do it and the mighty rocket delivered. Five engines and seven and a half million pounds of thrust to get that monster off the launch pad and yet it was not that enormous concentration of power that freed Apollo 8 from Earth's gravitational pull but Saturn's third stage, the S-IVB, which ultimately put the spacecraft into Earth orbit

and then later powered them out of Earth's influence and on to the Moon. Remember, in Earth orbit a spacecraft, and that includes all those Mercury and Gemini flights, is cruising along at right around 18,000 miles an hour, so the move up to nearly 25,000 mph was really not that much more. The single-engine S-IVB delivered. You might take note of our Apollo artistry where you'll clearly see the S-IVB third stage just above that tapered cone atop the first two stages. Not just the U.S., but also the world was about to get a Christmas week treat that would leave millions shaking their heads in awe and admiration. There would be television and music and voices from space that many of us will never forget.

The race to the Moon took two days and the crew appeared on live television on each of those days around mid-afternoon. On Christmas eve as Apollo 8 neared the Moon, they turned their Command and Service Module complex around so that the SPS (Service Propulsion System) engine was facing forward. After they disappeared behind the Moon where no man had ever gone before (and were out of voice contact with Houston and the world as well) they fired that SPS engine to slow themselves down and slide into a nearly circular orbit just about 70 miles above the Moon. Previously they had been in an elliptical orbit but now they were ready for some serious observations and picture taking for future Apollo missions.

So it was, that for the next 20 hours and ten lunar orbits the men were kept busy with their cameras and sextant, taking photos of prospective Apollo landing sites and gathering bearings for those proposed landings. And they still found the time to appear on television again.

They were kept in touch with an entire world through a network of land stations, instrumented ships at sea and instrumented aircraft plus all those cable, telephone, teletype and radio facilities across the globe. Except for their worrisome time behind the Moon, they were in constant touch with us Earth folk.

If you were listening and watching the mission of Apollo 8 and somewhat prone to "goose pimples," as we like to call that feeling when emotions cause the hairs on our arms to stand up, then you will never forget Christmas eve from space when the crew took turns

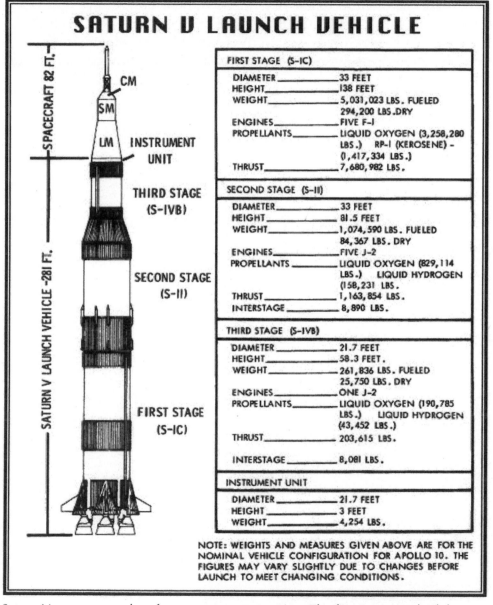

SATURN V LAUNCH VEHICLE

SPACECRAFT 82 FT.

CM

SM

LM   INSTRUMENT
      UNIT

THIRD STAGE
(S-IVB)

SATURN V LAUNCH VEHICLE -281 FT.

SECOND STAGE
(S-II)

FIRST STAGE
(S-IC)

FIRST STAGE (S-IC)	
DIAMETER	33 FEET
HEIGHT	138 FEET
WEIGHT	5,031,023 LBS. FUELED 294,200 LBS.DRY
ENGINES	FIVE F-1
PROPELLANTS	LIQUID OXYGEN (3,258,280 LBS.) RP-1 (KEROSENE) - (1,417,334 LBS.)
THRUST	7,680,982 LBS.

SECOND STAGE (S-II)	
DIAMETER	33 FEET
HEIGHT	81.5 FEET
WEIGHT	1,074,590 LBS. FUELED 84,367 LBS. DRY
ENGINES	FIVE J-2
PROPELLANTS	LIQUID OXYGEN (829,114 LBS.) LIQUID HYDROGEN (158,231 LBS.
THRUST	1,163,854 LBS.
INTERSTAGE	8,890 LBS.

THIRD STAGE (S-IVB)	
DIAMETER	21.7 FEET
HEIGHT	58.3 FEET.
WEIGHT	261,836 LBS. FUELED 25,750 LBS. DRY
ENGINES	ONE J-2
PROPELLANTS	LIQUID OXYGEN (190,785 LBS.) LIQUID HYDROGEN (43,452 LBS.)
THRUST	203,615 LBS.
INTERSTAGE	8,081 LBS.

INSTRUMENT UNIT	
DIAMETER	21.7 FEET
HEIGHT	3 FEET
WEIGHT	4,254 LBS.

NOTE: WEIGHTS AND MEASURES GIVEN ABOVE ARE FOR THE NOMINAL VEHICLE CONFIGURATION FOR APOLLO 10. THE FIGURES MAY VARY SLIGHTLY DUE TO CHANGES BEFORE LAUNCH TO MEET CHANGING CONDITIONS.

Saturn V was a poster boy for corporate cooperation. The first stage was built by Boeing, the second stage by Rockwell and the third stage by McDonnell Douglas giving that company a role in all three stages of NASA's pathway to the Moon.

reading the ten verses of Genesis, the Story of Creation. Borman punctuated this unforgettable time with a sign off, "God bless all of you on the good Earth."

They were never far away from Christmas comments even when, in only the second hour of Christmas day, they fired that indispensable SPS engine to escape lunar orbit and head for home. It worked, of course, as it always would, but it still inspired Jim Lovell to say, "Please be informed…. there is a Santa Claus."

There were more television shows on the way back to Earth but then it was time to stow gear and pay close attention to the details of re-entry as Apollo 8 once again approached 25,000 miles an hour. Our network anchors and their space specialist cronies would repeatedly stress that even the slightest miscalculation when attempting re-entry into the Earth's atmosphere could result in either a burn-up or skipping like a flat rock on a pond on out into space and eternity. Nothing like building suspense in your audience but they were not exaggerating. There was plenty of danger there. It was time to "thread the needle" and hit the so called "window" that we were hearing so much about at both launches and re-entries.

Re-entry was affected with the blunt or heat shield end of the spacecraft first, of course, and again we learned that temperatures on that shield could rise to several thousand degrees as Earth's friction took its toll and g-forces on the astronauts rose dramatically. They had all felt those forces through thousands of training hours in Houston.

Imagine a spacecraft re-entering the Earth's atmosphere at a speed of close to 25,000 miles an hour and slowing to around 300 mph by the time it was about 5 miles above the Earth. It gives you a good picture of why those temperatures get so high on the heat shield. At 24,000 feet, the small drogue parachutes deploy to slow the Command Module down enough so the big orange and white canopies don't get torn to shred when they deploy at 10,000. The "big three" then slow the descent rate to 27 feet per second, enabling a survivable, if not crew-friendly, splashdown.

At that point, if the recovery carrier and any other ships of the recovery force are in the area, they get an up-front show that only they will have seen in person. As it was, the prime recovery carrier USS Yorktown, one of the Navy's most legendary flattops, was ridiculously close BUT it was 4:50 in the morning on December 27.

There were lots of ship's crew on deck but they didn't see much in the dark. With the UDT guys (the frogmen) doing their usual professional jobs, they had Borman, Lovell and Anders up into one of the recovery helicopters and on the deck of the Yorktown in an hour and 20 minutes and, by then, the sun was up.

Yorktown's CO was Captain J. G. Fifield. Helicopter Squadron 4 (The Black Knights) was on deck for the first of its five missions and the frogmen were from UDT-12. No, the crew of Apollo 8 did not land on the Moon, but they were the first humans to orbit it and, of course, see what the back of the Moon looked like. It was different than the side we see all the time and they had the pictures to prove it. They had taken part in the first manned launch of the Saturn V from Complex 39, and were the first men to escape from Earth's gravity. No man had ever gone so far from Earth or so fast or had communicated over the lunar distance or appeared on TV from the Moon's vicinity. They showed us "Earth rise" for the first time. Everything they did fell into that rare category – a first.

So it was that when Jim Lovell headed for the Moon in the edge-of-your-seat hair-raising flight of Apollo 13 in April of 1970 and was lucky to get back alive without having landed on the Moon, he was not being facetious when he said, "I would rather have been on Apollo 8 than Apollo 11. It was the high point of my space career." Many others, including Neil Armstrong and other Apollo astronauts were inclined to agree.

Apollo 8 broke the bounds which tie us to this planet. It was a time of exhilaration and celebration in the same year that saw both Martin Luther King, Jr. and Senator Bobby Kennedy assassinated. It ended our year on a very high note when we needed it most.

Now where do we go to see this record breaker these days? Check out the Museum of Science and Industry in Chicago. It's there.

Earth rise. A sight seen and photographed by only a privileged handful of humans – the men who went to the Moon and back. This beautiful, peaceful picture was taken by the crew of Apollo 8.

USS Guadalcanal was again the recovery ship.

Just call me "Gumdrop." The astronauts were back in the business of naming their spacecraft without creating a lot of controversy and debate. Others would label Apollo 9 "an anti-climax" because it was not going to the Moon. Not as glamorous but just as important because this time the LM (Lunar Module) was going along, not just for a joyride, but some serious testing – docking and undocking.

Now we have said before, that the Saturn 1B became the launch vehicle for Earth orbital Apollo missions while Saturn V was rolled out for the lunar missions. True, but this mission broke that rule because it did carry all the equipment needed by a lunar mission. Therefore, despite remaining in Earth orbit, Saturn V had to be the launch vehicle; and since all upcoming Apollo missions would head for the Moon, Saturn 1B's career was shelved until the Skylab series began in May of 1973. The Skylab flights would all be in Earth orbit.

So it was a first after all. Who said "anti-climax?" The first manned flight with all the lunar hardware on board.... but...in Earth orbit.

Lift-off took place from Launch Complex 39-A as the clock struck 11 on the morning of March 3, 1969. It would last for just over ten days and 152 orbits of the Earth.

With all that hardware going along for testing, not only Mission Commander Jim McDivitt and Command Module Pilot Dave Scott would have major responsibilites. A primary mission objective was to have Lunar Module Pilot Rusty Schweickart test all aspects of the LM in Earth orbit. Oh yes, not only the Command Module had a name this time (Gumdrop) but so did the gangly-legged LM. They called it "Spider" and with one look you would probably agree.

At this point we direct your attention to the accompanying NASA illustration which clearly shows what a Saturn V lunar package was all about above the S-IVB third stage, the final 82 feet in other words.

The launch escape system would be the first piece of hardware to be jettisoned in an uneventful launch and this mission was one of them. So now we are down to the Command Module with its protective cover sitting atop the Service Module which, in a lunar mission, would provide the propulsion for insertion into orbit around the Moon and provide the same power for the CM's return trip. Just below that is what we might call the Lunar Module garage. There sits the "Spider" protected by the adaptor panels which would soon be jettisoned.

At two hours and 41 minutes into the Apollo 9 mission the S-IVB was separated from the Command and Service Modules. That is when those adaptor panels were thrown away, leaving the LM sitting atop S-IVB and still physically attached to it. Now the CSM turned

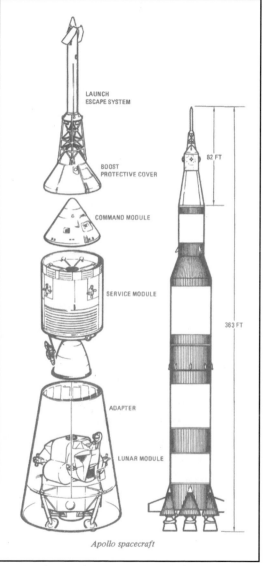

Apollo spacecraft

NASA chart – from Lunar Module to the Launch Escape System. (Courtesy of NASA)

around and docked with the LM some 3 hours into the mission. Four hours after launch, S-IVB and CSM were finally separated, by about a one-minute burn, which kicked the S-IVB into a very long elliptical orbit called an escape trajectory. So long and thanks for everything.

March 4 passed with the crew conducting a series of pitch and roll, yaw maneuvers, tracking landmarks, and increasing the apogee of their orbit by firing the Service Propulsion engine.

The next day, with Dave Scott minding the Command Module, McDivitt and Schweickart entered the LM through the docking tunnel at the point where the two spacecraft were joined, nose-to-nose. They checked out all LM systems, did the first of two series of telecasts and fired the LM's descent propulsion system…the system that would guide and power the LM down to the surface of the Moon later that year. They then crawled back into the Command Module.

On March 6, it was back to the "Spider" for more live TV after which Schweickart took a 37 minute space walk, standing on what was called the LM's porch and taking a bunch of pictures while cruising around with the new portable life support system and EVA mobility unit. While Schweickart was doing that, Dave Scott was doing his own standing EVA from one of the Command Module hatches so you can just picture them taking pictures of each other.

A day later and it was time to get a bit bolder. Separate the LM from the CSM and use the thrusters to put some distance between the two, the idea being to prove that they could separate and get back together repeatedly if necessary. This they did and would do again numerous times throughout the 10-day mission. The Moon was looking closer all the time. As a matter of fact, with 140 hours left in the mission, the crew had already racked up success with 90% of their objectives. The dockings and undockings were going smoothly with only one instance where docking latches hung up and would not allow separation. Like most problems, it was eventually overcome.

More EVA's, television broadcasts and picture taking.

In the world of space and space travel, dollar figures often soar higher than the machines they help create. When you think of how much it had to cost to have Grumman build those spindly-legged Lunar Modules out in Bethpage, Long Island, New York, and then realize that Apollo 9, after using one for docking practice for over a week, simply cut it loose and sent it off on another of those deep space trajectories such as the route taken to oblivion by the S-IVB, it amounts to what car buyers like to call "sticker shock." Actually it was ground control who sent the signal that sent the Lunar Module away. The crew watched it fade away for a while before getting down to checking systems, doing those navigational exercises and taking still more photographs.

One record that was consulted for this account put the length of the mission at 6 million kilometers. That's a lot of miles no matter how you calculate it. When it was over, Apollo 9 splashed down in the Atlantic at 12:54 p.m. EST northeast of Puerto Rico, about 180 miles east of the Bahamas and most comforting of all – within sight of the USS Guadalcanal. The skipper was Captain R. M. Sudduth. The helicopters belonged to HS-3 and the Recovery chopper pilot was Commander G.M. Rankin. The frogmen were from UDT-22. Just in case you're running a tab on this adventure – it cost an estimated 340 million. The contented crew of Apollo 9 was aboard the carrier less than an hour after splashdown.

Apollo 9…. now playing (but actually just sitting there) at the Michigan Space and Science Center at Jackson Community College in Jackson, Michigan. With the exceptions of the Johnson Space Center in Houston, the Smithsonian in Washington and facilities down around Cape Canaveral, which all display more than one spacecraft, you will see that NASA has done a fine job of spreading these historic objects around, not just in the U.S., but in other countries as well.

Russell "Rusty" Schweikart, became the second civilian to fly in the U.S. space program, Neil Armstrong having been the first.

USS PRINCETON (LPH-5)

The USS Princeton was the primary recovery ship for Apollo 10 and its crew of Tom Stafford, John Young and Eugene Cernan. It was Ticonderoga class – 888 feet in length.

A BRIEF HISTORY OF THE USS PRINCETON (LPH-5)

The USS Princeton (LPH-5) probably went through more evolutions than any other carrier in her time. When her keel was laid down at the Philadelphia Navy Yard on September 14, 1943, she was supposed to launch as the Valley Forge; but, as happened to a number of other ships built during wartime, she got a name change before sliding down the ways. She became the USS Princeton (CV-37) on November 21, 1944, was launched on July 8, 1945, and commissioned on November 18 of that year.

Now comes the part that frequently confuses and we've been down this road before. Designed as an Essex class carrier, the Princeton was Ticonderoga class after numerous modifications, the most obvious of which was the reshaping of her bow into a rather elegant "clipper" form that created deck space for two 40 mm quadruple gun mountings which greatly improved the ship's forward air defenses. There were other differences, most of them unnoticeable, such as massively revised and improved ventilation systems, upgraded radar and so on.

The one thing that Essex class carriers, which became Ticonderoga class, had in common was length – 888 feet and they came to be referred to as the Essex class "long hull" carriers. Six carriers modified into Ticonderoga class became primary recovery ships for manned space flights and Princeton was one of them.

While Princeton missed out on World War II, she did spend her first few years in the Pacific starting in June of 1946 and up until entering into the Pacific Reserve Fleet on June 20, 1948, but her retirement was to be short-lived.

(Navy photo)

War, or as some called it, "police action," broke out in Korea just fifteen months later; and by December of 1950, with a crew of intensively trained reservists, she was part of Task Force 77 and her planes and pilots (Air Group 19) helped re-institute jet combat air patrols over Korea's battle zones. They flew hundreds of sorties and flew in support of Marines fighting their way down the bitterly cold roads from Chosin Reservoir (the troops called it "Frozen Chosin") to Hungnam.

Princeton's planes were in the air constantly through the first half of 1951 but she was back in San Diego for overhaul by late August. Eight months later Princeton rejoined TF 77 and again her planes flew against Korean targets for 138 days.

She was re-classified CVA-37 (attack carrier) on October 1, 1952, got a two month respite in California, but was back in Korean waters by February of 1953. She was there when the truce was achieved on July 27 but headed for San Diego in September.

The following January came the USS Princeton's latest re-classification (but not her last) when, after conversion at Bremerton, she became CVS-37 and that meant anti-submarine operations which kept her occupied for the next five years.

Although seemingly slated for decommissioning after that evolution, the Princeton was instead redesignated LPH-5 (her last change) in March of 1959 and spent the rest of her naval career as an amphibious assault ship. She never did get an angled deck like so many of her sister carriers in ASW work. It was at this point that Vietnam entered the picture and Princeton was heavily involved in that conflict as an LPH with her helicopters ferrying Marines and Army units into the fighting. Princeton remained in that basic assignment almost to the end of the 1960's, but in April of 1969 she got a real change of pace when she was named primary recovery ship for the Apollo 10 lunar (non-landing) mission. Details on that upcoming.

In January 1970, Princeton was decommissioned and stricken from the Naval Vessel Register. She was sold for scrapping in May 1971.

After two and a half decades of service, the USS Princeton retreated into Naval history with 8 battle stars for the Korean conflict.

THE FLIGHT OF APOLLO 10 WITH ASTRONAUTS GENE CERNAN, JOHN YOUNG AND TOM STAFFORD FROM MAY 18 TO 26, 1969

While the news media sometimes appeared pre-occupied with putting labels on the space missions and classifying them as "firsts" or "anti-climaxes" for one reason or another, the crews themselves never let on that they were anything less than eager and honored to do their part in the program. The Apollo 10 mission was no exception. So what if it was not the first to go to and around the Moon? It was, however, the first to make the trip with a fully configured Lunar Module – all the bells and whistles required for an actual landing. Besides, two out of three of Apollo 10's crew would go on to actually walk on the Moon in later missions. Oh yes, lest we forget. Nicknames. The CSM combo got the name "Charlie Brown" and you might almost guess that the LM would end up being called "Snoopy."

This would be one of the few times during manned missions with splashdowns in either the Atlantic or the Pacific that a recovery carrier would journey out on its lone assignment in the program. In this case it was the USS Princeton. Essex, prime recovery ship for Apollo 7, was another example.

As it turned out, Apollo 10 truly was the dress rehearsal for mankind's first lunar landing but it was not necessarily cast in stone at the time of lift-off. If things had gone terribly wrong and serious questions raised, then NASA would not have felt all was ready for the big one. But objectives were being steadily attained and marked off as this program evolved, and Apollo 10 would fit that mold and set the table for space history to be made two months later.

Launch time was 11:49 a.m., on May 18, at the Cape and there had been no delays. Nice way to start. Just as a guideline as to the time required to reach the succeeding stages of a lunar mission, we will use Apollo 10 as an example, although not all missions to follow would be exactly the same with regard to time spent in Earth orbit before heading for the Moon.

In this case Apollo 10 achieved Earth orbit at around 12 minutes into the mission and was on its way out of that orbit and on to the Moon at about 2 hours, 40 minutes from lift-off. Here is where things take a little longer. Lunar orbit insertion would not be achieved for nearly 76 hours since leaving Launch Pad 39A.

Just about a three-day cruise, in other words, before the spacecraft entered its initial elliptical orbit and after some tweaking and firing of thrusters, rounded it out to a nice, manageable circular orbit for the very important LM usage coming up. On May 22, Tom Stafford and Gene Cernan entered the LM and fired the reaction control thruster on the Service Module to separate their new home from the CSM. If that sounds a trifle confusing, remember that the CM itself has no rocket propulsion engine. The power of separation came from the SM engine. If that sounds like the Command Module is very dependent on other elements of the launch vehicle, you're right. It needs a push now and then from a separate source.

The LM was taken down to an altitude of just about 9 miles above the lunar surface where some serious picture taking was done

including nice close-ups of the Sea of Tranquility soon to become a by-word for the first landing site on the Moon. They also got in some nice quality live color television for us Earthlings.

On every mission, medical testing became on ongoing essential with in-flight medical monitoring telling NASA physicians exactly how hearts were beating and men were breathing in the spacecraft environment. Pre- and post-mission testing was also a must. In the near weightlessness of space there would be loss of bone density and this had to be watched closely – all the time. There was bone density loss and it was expected, but it was always within acceptable limits. It was a situation that corrected itself with time, back on Earth.

On May 23 the LM and CSM got back together for the last time. Late that day the LM was sent rocketing into a solar orbit, and on May 24 after 31 orbits of the Moon, the rockets on the CSM were fired to bring Apollo 10 home again. The carrier Princeton, with Captain Carl Cruse, was waiting 400 miles east of American Samoa when, on the 26th of May, 1969, Apollo 10 splashed down just 3.4 miles from her deck. Officers and crew could easily watch HS-4 Recovery helicopter pilot Chuck Smiley move in and pick up the astronauts, ably assisted by the frogmen of UDT-11.

The space community was getting ready for some Moon rocks and anything else the first Moon Walkers might soon find there.

As for Apollo 10, it now resides at the Space Museum in London, only the third time an American spacecraft has left these shores.

USS HORNET (CVS-12)

Primary recovery ship for the first and second lunar landing missions in 1969. Apollo 11 with astronauts Neil Armstrong, Buzz Aldrin and Mike Collins and Apollo 12 with astronauts Pete Conrad, Dick Gordon and Alan Bean.

The USS Hornet was an Essex class carrier – 872 feet in length - one of the "short hull" crowd. Some remained straight-decked but Hornet was given the angled deck that brought with it the designation CVS and put her into the anti-submarine warfare world.

A BRIEF HISTORY OF THE USS HORNET (CVS-12)

The Hornet, which recovered America's first lunar landing crew on July 24, 1969, was the eighth U.S. ship to bear that proud name.

It must be mentioned that Hornet number 7 was the carrier that sent a brief but surprising message to Japan on the morning of April 18, 1942. On that day, Lt. Colonel Jimmy Doolittle's and his hardy band of 70 officers and 64 enlisted men climbed aboard their 16 Army B-25 bombers, lumbered off the Hornet's deck in heaving seas and delivered a startling strike on Tokyo. The book *Thirty Seconds Over Tokyo* and the movie of the same name, which followed, were not exaggerations. The surprise raid was over quickly, and, as most of you probably know, not all those planes made it safely back to so-called "friendly territory" – namely, mainland China. They had been launched prematurely because of detection of the carrier by a Japanese patrol boat. Consequently, none of Doolittle's planes had enough fuel to get back to that ship; but the Colonel and most of his men did survive and, while doing minimal damage to the Tokyo area, they did wonders for American morale so early in the war.

That Hornet went through the battle of Midway the following June but reached the end of her gallant life during the battle for Guadalcanal in October of that year. Japanese carrier-based planes delivered fatal blows off Santa Cruz Island in the Solomon group, and Japanese destroyers finally sent Hornet (CV-8) to the bottom on the 27th of October.

By early 1943, an Essex class carrier was taking shape at the Newport News Shipbuilding and Dry Dock Company in Newport News, Virginia. The Navy decided that it would become Hornet (CV-12), the eighth ship to bear this name.

When she was launched her length was 872 feet. By the time we boarded her at Pier Bravo in Pearl Harbor in July of 1969, the handout given to all members of the news media as we came aboard told us she was then 894 feet in length and had gained considerable weight since her initial commissioning, which can only be attributed to a series of overhauls and refitting over the years.

Hornet (CV-12) was launched on August 30, 1943, but there were sea trials and shakedown cruises to get under her belt before she joined the famed Fast Carrier Task Force 58 in the Marshall Islands on March 20, 1944. Hornet proceeded to lend her air support to protecting the invasion beaches in New Guinea…conducted massive aerial raids on Japanese bases in the Caroline Islands and, commencing in early June, launched raids on Tinian and Saipan in the Marianas as well as heavy bombing attacks on Guam and Rota.

It was during these operations that Hornet was in the thick of what became known as "The Great Marianas Turkey Shoot." The Hornet and other carriers combined to down some 315 Japanese land-based and carrier planes in a single day.

The cost to the Japanese Naval and air forces grew even worse when, during the Battle of the Philippine Sea, Hornet's planes attacked the Japanese carrier Zuikaku, causing that ship and her companion carriers to lose nearly 400 of their 430 planes.

She went on to support landings on Morotai, Leyte, Iwo Jima and Okinawa before joining with other carriers of Task Force 58 to attack and sink Japan's super-battleship Yamato. Throughout April the Hornet fought off desperate Japanese kamikaze attacks and when the smoke had cleared, she could take credit for over 1,400 enemy aircraft downed and more than a million tons of enemy shipping sent to the bottom of the Pacific.

For all her accomplishments in battle, the USS Hornet received the Presidential Unit Citation and 9 battle stars. Though attacked 59 times, Hornet suffered no major battle damage, although a typhoon during the battle of Okinawa tore up the front of her flight deck.

Post-war duty included a role in "Operation Magic Carpet" – bringing our troops home from war zones.

Decommissioned in January 1947, she was recommissioned in March 1951, but went in for 27 months of conversion into an attack aircraft carrier and her designation became CVA-12.

In December, 1955, the Hornet entered Puget Sound Naval Shipyard in Washington for extensive modernization. The new hurricane bow and angled flight deck, which many Essex class carriers received, facilitated the simultaneous launching and landing of aircraft. Later, she was slated for the conversion to a full-blown anti-submarine warfare carrier and her final designation, CVS-12, which came about in the first half of 1958.

No longer were enemy ships or planes trading shots with Hornet but she spent the following years carrying out operations with the 7th Fleet ranging from the coast of Vietnam to the shores of Japan. On August 25, 1966, she was baptized into the spacecraft recovery business when she retrieved the unmanned suborbital Apollo flight AS-202 near Okinawa.

For her next to last space-related effort she was picked as prime recovery ship for our first manned mission to the Moon – Apollo 11 in 1969. It turned out to be a nice year for this storied carrier as Hornet continued on as prime recovery ship for our very next mission to the Moon – Apollo 12 in mid-November of that same year.

The clock, however, was definitely running on this veteran of World War II, and the Hornet was decommissioned for the last time on June 26, 1970. After years of inactivity, struck from the Naval Vessel Register on July 25, 1989, and sold to be scrapped. That was the fate of most of the fine ships in this book but, considering her history and especially the significance of Apollo 11, it is a pleasure to note a group of historically minded citizens stepped forward and got her declared a National Historic Landmark (#1029). Hornet was eventually donated by the Navy to the Aircraft Carrier Hornet Foundation, who opened her to the public as a museum in 1998. She is currently at Pier 3, Alameda Point (the former Alameda NAS), less than 30 minutes from downtown San Francisco.

THE FLIGHT OF APOLLO 11 WITH ASTRONAUTS NEIL ARMSTRONG, BUZZ ALDRIN AND MIKE COLLINS FROM JULY 16- 24, 1969

One Giant Leap for Mankind. First human steps on another world. Mankind breaks the bonds and for perhaps more than any other time in human history, the audience was truly worldwide and, with few exceptions, literally billions of cheerleaders.

For those of us in the news media and armored against the heart-breaking stories we cover almost as routine, this culmination of incredible talent, imagination and devotion to a cause, was the biggest breath of fresh air any of us could ever hope to witness –whether from the Cape, Houston or from the deck of a mighty and historic American aircraft carrier – the USS Hornet.

I must get a bit more personal in this mission re-telling simply because, after "pool reporter" assignments on three straight Gemini missions in 1966, I, as a correspondent for the Mutual Radio Network (no TV), had not seen another down-range assignment since then. Here it was midway through 1969 and the biggest adventure of our lifetimes was right in front of us.

It was sometime in the spring of that year that those network news execs who went to the meetings where "pool" assignments were drawn, decided wisely that instead of sending just two of us, as on previous missions, this time it would be a foursome. All three television networks were represented. Ron Nessen from NBC, Dallas Townsend from CBS, Keith McBee from ABC and yours truly as the lone reporter for radio – worldwide, on recovery day. Super-Bowl hype had nothing on Apollo 11. They had to send four of us out there. When Apollo 11 came home on July 24, my three colleagues shared the description for television on three open microphones while I did a solo act for radio from my favorite nook on the 06 level.

Most anyone awake on the morning of July 16 and old enough to care, watched and crossed their fingers, held their breath and prayed for Armstrong, Aldrin and Collins as they sat, their backs to the Earth, their eyes skyward atop man's mightiest rocket, the Saturn V.

Only later would we learn that on the morning of the launch, Dr. Thomas Paine, NASA's administrator (the Boss!) told the crew their safety was more important than anything else and that if anything looked wrong they were to abort the mission... get out of there. And if that should happen, the trio would be immediately assigned to the next landing attempt. Very reassuring words.

Apollo 11 lifted off from Launch Complex 39-A at 9:32 on the morning of July 16, 1969. There had been no delays.

Smoothly, they went through the separation of the Command Module Columbia from the Saturn rocket's third stage, turned around and connected nose-to-nose with the lunar module Eagle which had been stored, as you know by now, in the third stage. NASA could show the world how to pack efficiently for long trips. Buzz Aldrin noted with great satisfaction that the entire operation proceeded perfectly. It was just 14 hours after lift-off when the Apollo 11 crew covered their windows and made a valiant effort to get 8 or 9 hours sleep, but that task always proved difficult and the hours they did get were, at best, fitful.

---

*If you own a computer and search the web, you will find a delightful print-out of the comments of all three members of the Apollo 11 crew on their way to the Moon and on their way back. Not only that, you will be on your way to a treasure trove of space history and information. Obviously, by using the base signature and changing the mission numbers you will find more and more space history. It will seem endless. The three men admittedly felt like tourists as they high-tailed it to the Moon, watching the Earth retreat and the Moon loom larger with each passing hour. All would admit to having a lot of thoughts, misgivings, about their chances of getting to the Moon and home again. But as each of them mulled over the prospects for success, they also openly admired and recorded their march through each and every step of the flight.

Mike Collins, who would command the sympathies of millions when he had to simply orbit the Moon while his crew mates went Moon walking, could not get over the change they observed as the Moon he had spent so many years looking at – a quarter of a million miles away and, according to legend, made entirely of green cheese, was suddenly filling his window completely and Mother Earth had traded places, becoming that little object in the sky.

And then, there they were at the Moon, and swinging around its left side only 300 nautical miles from it, aware that it really IS a moving target and the geniuses who had put the whole space program together had gotten them there with what Collins called super-accurate predictions. All three expressed their appreciation repeatedly throughout the mission.

Apollo 11 was not loaded down with a lot of scientific gear, certainly not when compared to all later flights. The whole idea was to get to the Moon and get back safely with some souvenirs. Moon rocks. Eighty-six hours into the mission, and that included 5 and half times around the Moon itself, the three men settled down for their last sleep period before the landing. It was a good sleep although it only lasted six hours. No interruptions. Houston's wake-up call came 93 hours into the mission and then, for the next 8 hours, they were totally devoted to the descent down to the lunar surface.

By the time they swung around the Moon for the fourteenth time, the CM and the LM were separated and all three men properly suited up and ready for the big step. As the world knows, Mike Collins remained in the Command Module and cruised in a circular orbit about 70 miles above the Moon. There were no seats in the Lunar Module so Armstrong and Aldrin stood at the controls, held in place by straps with a certain amount of elasticity. Sort of like bungee cords.

The LM was rotated to a point where Neil Armstrong could see ahead and pick out what looked like a safe landing site. They were, of course, using photos taken by Apollo 10 and it was a tremendous benefit. Well into the 12 and a half-minute descent engine burn, they were being reminded of their fuel supply and there didn't appear to be a whole lot to go around. Houston was getting anxious. Armstrong was Mr. Cool.

Three and a half years later, Gene Cernan, last man to walk on the Moon (during Apollo 17) would comment that a lunar landing was easier than putting his fighter plane down on a carrier deck. Only a Navy pilot could contest that and only if he also had a chance to land on the Moon. At any rate, Armstrong was enjoying the sophisticated on-board computer in the spidery LM, which did a lot of the routine work of piloting and making it look as easy as Cernan would later maintain that it was.

There were a few flashed computer alarms on the way down but back in Mission Control, a man who knew that computer best, Steve Bales, quickly figured out that it was information overload and that the equipment was designed to handle it – which it did. As a result, he was able to confidently assure the two men in the LM that it was alright to continue. For his role in keeping the lunar landing on track, Bales was later able to stand alongside the Apollo 11 crew at the White House and receive a Presidential decoration for his efforts.

Buzz Aldrin was not just standing there like a tourist alongside Armstrong during the descent. He was very busy passing along computer output to Neil and was giving his LM partner an "angle," so-called, every few seconds. That rapid-fire information convinced Armstrong that their computer would take them down into a field of boulders. This is where, in mission after mission, the thousands of hours of flying experience of most of the astronauts would kick in and "seat of the pants" flying would take over. Armstrong, a former naval aviator with many carrier landings to his credit, simply switched to manual control and within seconds was handling the LM much like you would fly a helicopter. He slowed their rate of descent drastically.

Aldrin kept up his "chatter box" feed of data while Armstrong "fine-tuned" things and Mission Control grew increasingly worried about that fuel supply. The LM was just 75 feet above the Moon when fellow astronaut Charles Duke in Mission Control told them they had 60 seconds of fuel left. Buzz Aldrin was saying about the same thing; but by the time they did, the LM was already kicking up lunar dust and within moments, Eagle had landed and engine shutoff had occurred. Piece of cake.

Neil Armstrong radioed his historic, "Houston, Tranquility base here, the Eagle has landed," and almost before the celebrating had died down, the crew of the first vehicle (some would call it a contraption) to land on another world had begun getting ready for an immediate lift-off in the event of any sort of problem or emergency. That was job number one in this case. So there was work to be done and some rest to catch up on. No seats, so they actually would lie on the floor of the

LM to do that. Imagine being first on the Moon and not being able to zip right out the door, down those steps and onto the Moon itself…immediately. Now was the time for utmost caution and preparation. Therefore, it was fully 6 and a half hours after the landing before they opened the hatch and Neil Armstrong let himself down that ladder, facing it and proceeding slowly. Moments later came some of the most historic, celebrated, repeated words ever said by anyone, anytime or anywhere. "That's one small step for man…. one giant leap for mankind."

A fixed camera showed Neil descending and taking that historic first step. Not many minutes later his LM partner joined him on the grainy, dusty, strange lunar surface. They were tentative at first, both of them, as they should have been and were told to be. It was a strange material they were cautiously shuffling their boots in, and they did not go too far from their equally strange looking home. In somewhat less than two hours, the two men moved the television camera further from the LM for a wider view of the history-making scene, deployed a few scientific instruments including gear to detect seismic shocks on the Moon, and collected big rocks and small rocks… a little over 40 pounds in all.

There was just so much room in the Command Module and total weight was always a key factor, so anything they collected on the Moon became a trade-off with something they had brought with them. Consequently, the instruments left on the Moon, even an expensive camera, (then and on subsequent missions) were purposely left behind to keep the ride home in balance.

They deployed an American flag, got a bit frisky and ventured further away from the LM and showed the TV audience the thrill and the effort of walking, hopping and jumping on the Moon. Other missions to come would get friskier and venture further and even have wheels and a motor to do it with, but this was Moon landing number one and no one was ready to put this gigantic effort at risk.

Early on it was assumed, especially by NASA Science Director Bill Hess, that there would be an ALSEP (Apollo Lunar Science Experiment Package) on board, but it got sand-bagged in a sea of debate leading up to the mission and was, eventually, set aside. There were some science experiments aboard on Apollo 11, but, as we said earlier, not many.

One other point that should be made here. Buzz Aldrin tried his darndest to drive core tubes into the lunar surface to obtain soil samples below that grainy surface but was never able to drive anything more than 8 or 9 inches deep. The sub-surface was very, very compact. NASA wanted better evidence of lunar history. In later missions, a new battery-powered tool would emerge and become standard equipment on Moon missions. Developed and built by Black and Decker, it did penetrate down several feet and bring up samples in a tube. That drill was the forerunner of a home handyman's collection of battery-powered tools that has become a multi-billion dollar, worldwide industry. The space program did that a lot… a whole lot.

Other missions would spend many more hours on the lunar surface and explore entirely different landscapes and ascend mountains. Armstrong and Aldrin spent but a few hours, two hours and 31 minutes to be exact, before they climbed those steps back up into the LM and closed the hatch. Then they had those rock samples to stow, other gear to be left behind (so far we Americans are the only known lunar litterbugs) and a lengthy list of basically housekeeping chores to knock off before they could curl up on that luxurious Lunar Module floor (only kidding). They would later call that "rest period" almost a complete loss.

Houston left them alone for about seven hours. Then…Houston's wake-up call was followed by a quick breakfast and preparation for the first-ever blast-off from the Moon and back to re-rendezvous with lonely Mike Collins in the Apollo 11 Command and Service Module.

Considering that only twelve men have ever actually stepped out onto the lunar surface, Mike Collins and the other five Command Module pilots who had to bide their time in lunar orbit have never been heard to gripe about their role in those missions. After all, they are in a pretty exclusive club themselves…. lunar orbiters. Mike would get to watch as his teammates blasted off from the Moon and climbed back up to him like a puppet on a string. He would describe that approach as steady as a rock and down the centerline.

Hatch, probe and drogue, the hardware which enabled CM and LM to join and lock and enabled Armstrong and Aldrin to scramble from one to the other, stayed with the LM which was soon sent on its lonely road to oblivion. If creatures from any other world have stumbled upon the spidery-looking "works of Earth men" which we have dispatched into the endless void of space, they haven't said anything yet. Or perhaps they have hauled in one or two and plan to use them in an

Earth landing at a time to be announced. And how would they know we call it Earth?

Back to business. That simply wonderful, works-every-time, Service Module propulsion engine fired as usual and propelled the world's newest heroes back to Earth and a nifty splashdown in the South Pacific on July 24, 1969. The USS Hornet commanded by Captain Carl. J. Seiberlich was on duty and, as noted here earlier, I was one of the four broadcast newsmen on board and my job was to tell the entire radio world exactly what was going on. I had been on three previous recovery ships for Gemini 9, 10 and 11 and each one of them was a mere few miles or less from our decks. This time, I clearly recall looking up anxiously through broken cloud cover, straining to see Apollo 11, because all our sources and reports told us it had to be in our neighborhood, and at an altitude where those three huge orange-and-white parachutes should be popping out. I was only one of many, including the sailors all over Hornet's deck, who spotted a brief flash high overhead, and shouted, "There it is" or something equally as brilliant. After that, it was a steady sail to a spacecraft that landed over the horizon…probably ten or twelve miles ahead of us. Upon landing, Apollo 11's trio was upside down in what NASA terms "stable two" position. You and I would call it very "unstable." In weightless, or near weightless space, being upside down was incidental and your body never knew it; but on the rolling South Pacific it was a different story. Always thinking ahead, NASA engineers equipped all spacecraft with balloons in containers packed in the nose section, which could be, at the touch of a button, inflated and force the spacecraft to pop up to the upright mode. It worked for the men of Apollo 11 and I know they appreciated that simple device.

Then the frogmen (UDT-11 and 12), leaped into the Pacific from the HS-4 helo, attached a sea anchor, the flotation collar, attached a communication cable to the spacecraft to talk to the crew, knocked on the door to ask if they were ready, and tossed in the BIG suits. These Biological Isolation Garments, which the frogmen also had to wear, were designed to protect themselves and us from each other, but especially us from any possible "Moon germs." Some in NASA ridiculed these precautions but, obviously, other heads prevailed and the BIG suit lasted for the first several missions until all concerned were convinced that the tranquil Moon bore us no malice or heretofore-unknown germs.

The frogmen wore suits with masks that filtered the air coming in to them while the astronauts donned BIG suits with masks, which filtered the air they were breathing out just in case they were, themselves, contaminated with something. While these lightweight, light-colored garments were being put on in the spacecraft, Lt. Clancy Hatleberg, the BIG swimmer, was dousing the entire Apollo 11 Command Module with Betadine, an iodine-smelling surgical wash. He then scrubbed each of the astronauts as they exited the spacecraft into the attached raft.

When all that precautionary stuff had been done, the suited astronauts were hoisted up to the hovering Sea King helicopter, piloted by Don Jones, and whisked away to one of Hornet's outside elevators and hurried into their home for the next three days, a sleek modified Airstream wheel-less trailer known as the MQF (Mobile Quarantine Facility). They were joined by NASA flight surgeon Dr. William Carpentier and MQF technician John Hirasaki while the space agency decided that they had not brought back anything we couldn't cope with from the Moon. They had not. But NASA being NASA, and redundant being the agency's middle name, we had not one but two MQF's end-to-end on the Hornet's hangar, just in case one malfunctioned in some manner.

Much has been said and written about a 21-day quarantine period, which began with their departure from the Moon. Some have assumed that the entire period was spent inside the MQF, which would have really been a cramped existence. Instead, the crew was transferred after three days from the trailer into a larger, but still sterile environment: the Lunar Receiving Laboratory at the Johnson Space Center in Houston.

As I was doing my radio-to-the-world broadcasting from twilight to mid-morning on the 24th of July, the Presidential helicopter came into dark view bringing Richard Milhouse Nixon to Hornet's deck and an immediate greeting by Hornet's skipper, Captain Seiberlich, and Admiral John McCain, CINCPAC, Commander-in-Chief-Pacific, and the father of now Senator John McCain, who was even then a prisoner-of-war in Vietnam.

Admiral McCain was a small, cigar-smoking dynamo of a man and we of the news media had the pleasure of meeting and talking with him at a gathering in the Admiral's very nice quarters a day or so before the splashdown.

President Nixon entered the hangar bay and manned a microphone in front of the MQF as soon as the Apollo 11 crew had been given initial medical check-ups. They settled down in front of the window-end of the trailer and exchanged pleasantries with their Commander-In-Chief. Then, quickly, the President of the United States, said his good-byes, re-entered the White House helicopter and headed back to super-secret Johnston Island.

Mission accomplished. The astronauts and all of us, I believe, were ready to collapse and log in a good night's sleep. A few hundred miles later the Hornet was back at Pier Bravo in Pearl Harbor. The MQF with the astronauts and two NASA folks, was off-loaded, and after a nice welcoming ceremony on the pier, loaded into an Air Force cargo plane and flown back to Houston.

Apollo equipment would later be transferred to other government agencies, donated to museums or sold off. The largest recipient was the

National Air and Space Museum of the Smithsonian Institute. Of the 4 MQFs built, only 2 survived and are on public display. The MQF and flotation collars for Apollo 11 can be seen at the new Udvar-Hazy facility at Dulles Airport.

I'll give you three guesses as to where you might find Command Module Columbia these days. It would be rather shocking if it ended up anywhere but the fabulous National Air and Space Museum at the Smithsonian in Washington, as indeed it did. It is very near Charles A. Lindbergh's Spirit of St. Louis…as indeed it should be.

THE FLIGHT OF APOLLO 12 WITH ASTRONAUTS PETE CONRAD, DICK GORDON AND ALAN BEAN FROM NOVEMBER 14 TO 24, 1969

USS Hornet was again the recovery ship.

You're in good hands with ALSEP. Sorry, I couldn't resist that. After Apollo 11 had to leave that significant scientific package behind for reasons ranging from the extra weight it represented, the space it required, and so forth, the fact that Apollo 12, the Yankee Clipper, left Earth with a complete ALSEP package on board was just one item, which immediately separated this mission from its predecessor. You will see what I mean shortly.

Apollo 12 lifted-off the pad at Launch Complex 39-A at 11:22 on the morning of November 14. Perfectly normal lift-off – right? Wrong! How about lightning striking the huge Saturn V as it was ascending, just about tripping all the circuit breakers in the Command Module. But the "All Navy crew" was intrepid (fearless, dauntless, according to Webster's Dictionary) and why not? The name of the Lunar Module was Intrepid. Of course, Mission Control was flipping out; but, in a just a few minutes, the crew had everything back to normal and no

permanent damage was ever uncovered. The rest of the flight was uneventful.

This mission was planned to land in an area where all concerned hoped and expected the men of Apollo 12 would find the unmanned Surveyor III, which had landed in a large crater known as the Ocean of Storms on April 19, 1967. Surveyor III was camera-equipped and immediately began sending back good pictures, which had a lot to do with NASA deciding to put one of its manned landings there. The original name given to the area quickly took a back seat to the name "Surveyor Crater" and it stuck. It was part of a series of craters…a cluster that looked a bit like a snowman with the Surveyor Crater making up what would be a snowman's mid-section.

Neil Armstrong was a business-like man, fully aware that the entire world was watching and he was determined to maintain a certain decorum and dignity even while trying to figure out how the first man on the Moon was supposed to walk and talk. Mr. Straight Arrow - much like John Glenn. Charles Conrad was a Command Module pilot who was quite the opposite. Relieved of the pressure of being first on the Moon, Conrad's natural tendency to laugh and have fun at most everything took over and gave its own tone to this mission. He never hesitated to rave about the view, the rocks, the other machine they landed so near – everything, and in Alan Bean he had a willing partner.

This mission was planned to be an exercise in precision targeting. Intrepid, the LM, was brought nearly all the way to the Moon's surface automatically – by radar and computer. Now, in about the last 400 feet, it was Alan Bean feeding the numbers to Pete Conrad who had switched over to manual control. All Moon landers learned the same lesson. As they got within a few yards of the dusty lunar surface, the chosen landing site would become obscured in much the same way a helicopter kicks up a storm of dust and dirt as it hovers near the ground. Consequently these landings might have been made with a few fingers crossed in those last seconds except the bulky gloves on the space suits wouldn't have allowed it.

They had the ALSEP (Apollo Lunar Science Experiment Package) kit with them this time and during the first of the two four-hour EVA's, setting up ALSEP's various experiments was at the top of their lunar surface assignments.

First they had to get down off that ladder. The bottom rung left a larger than normal distance between itself and the landing pad, which sat in the lunar dust. As Conrad, a short guy, took that last jump down he felt compelled to say, "Man, that may have been one small step for Neil, but that's a long one for me." Alan Bean was not far behind Pete. When they both got oriented, they got down to the business of setting up a TV camera, a seismometer, a laser reflector, a solar-wind analyzer and a number of other scientific packages which did not make it onto Apollo 11.

Another first. This mission brought with it a small, very carefully packaged, electric generator run by a small decaying plutonium source. Now you know why it was carefully packaged. Hot stuff. But it put out a steady 75 watts of power to all those gadgets they had deployed around them.

Unfortunately, one of the few glitches on this mission turned out to be the color TV camera. Oops. While Alan Bean was attempting to move it away from the LM, he accidentally pointed it at a bright reflection off the LM and it burned out the imaging tube in the camera. Understandably, if not regrettably, the television audience and the networks lost interest soon after that camera bit the dust. Nonetheless, Conrad and Bean had that formidable 70 mm Hasselblad with them and this was no time to spare the film. They took many pictures of the Surveyor III, resting only a short distance from their landing spot, but not scheduled for a real close-up inspection until their second EVA.

That first four-hour EVA was a thing of joy and they did not hold back their exuberance. When it was time to get back into Intrepid and try to rest up, they removed helmets and gloves but not those big bulky suits. It simply took a long time to get in and out of them and they did not want to rob themselves of their time out on the lunar surface. Apollo 11's astronauts had just the floor of the LM to sleep on. These fellows had hammocks and a heater, but despite logging five quiet hours, both men would admit they had hardly slept at all. Then it was time for a quick meal, those clever dried and re-constituted entrees and tubed delights. Much of these necessities were developed by the Stouffer Company and, on a printed menu at any rate, sounded pretty good.

By the time Pete and Alan ventured out for EVA number two, they had learned enough new tricks to burn up a lot less energy accomplishing their tasks of picking up rocks, taking pictures and ranging a lot further away from the LM than they did the first time. They found bending over very laborious, and probably due to their experience and post-mission comments, future astronauts would wear space suits with more flexible waists.

During the real close-up inspection of Surveyor III, they saw how the spacecraft had actually bounced upon its initial landing and skidded down the crater a short distance. Fearful that it might even slide while they were there, they did not approach it from below but from the side. It looked light brown but had been very white when launched back in 1967 and lunar soil was definitely not brown. What was the story? It was dust covered and in thin enough layers, and with the right background, lunar soil looks brown. Mystery solved.

Surveyor III had been equipped with a small-motorized shovel for digging into the lunar surface and televising that event. Conrad and Bean now had the chance to see first hand if the trench cut by Surveyor had changed – caved in, perhaps. As far as they could tell, there was nary a grain of lunar soil out of place. Small wonder then, their footprints are up there for eternity – unchangeable along with the imprints left by ten others.

What a shame the TV camera was knocked out of commission so early in this mission. They accomplished so much but nobody saw it as they did. They returned home with about 75 pounds of Moon rocks and soil. After rejoining Dick Gordon in the Command Module, they sent the LM's ascent stage back into the Moon. Crash! The resulting Moonquake registered for over an hour on the seismometers they left behind.

Splashdown occurred a few minutes before 4 p.m. on November 24 in the South Pacific. The carrier Hornet, still commanded by Captain Sieberlich, was on station. Antisubmarine Warfare Squadron HS-4 again provided the helos (its fourth mission) with Commander Warren Aut at the controls of the tried and true #66. The UDT-13 group, with Lt. Robertson in charge and Ernie Jahnke in the "decon" swimmer role, perfomed flawlessly.

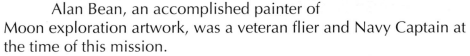

Apollo 12 can be seen at the Virginia Air & Space Center in Hampton, Virginia, which is just north of Norfolk.

Alan Bean, an accomplished painter of Moon exploration artwork, was a veteran flier and Navy Captain at the time of this mission.

USS IWO JIMA (LPH-2)

The USS Iwo Jima was the primary recovery ship for the troubled Apollo 13 mission from April 11 to 17, 1971, with astronauts Jim Lovell, Fred Haise Jr. and John L. Swigert.

Like the Essex, when you are first in line, they name a whole class of ships after you. Iwo Jima, therefore, as the first ship designed from the keel up as an amphibious assault ship, gave her classification name to all those that followed. These ships did launch with minor differences. Guadalcanal is listed in one historical record as exactly 10 feet shorter than Iwo Jima but apparently equipped to accommodate 157 more combat-ready Marines than Iwo Jima. It took a trained eye to tell them apart.

A BRIEF HISTORY OF THE USS IWO JIMA (LPH-2)

Life began at the Puget Sound Naval Shipyard in Bremerton, Washington, where the Iwo Jima was launched on September 17, 1960. Commissioning occurred on August 26, 1961.

Since Iwo Jima was the first of the breed, it might be useful here to go into a little more detail on the mission designed into these ships. Website communicator Bruce Sharpe writes, "Modern U.S. Navy Amphibious Assault Ships are called upon to perform as primary landing ships for major portions of a Marine Expeditionary Force and Marine Expeditionary Brigade. These ships use Landing Craft Air Cushion, conventional landing craft and helicopters, to move Marine assault forces ashore. In a secondary role using AV-8B Harrier aircraft (the so-called "jump jets" which take off straight up and then fly like conventional aircraft) and anti-submarine warfare helicopters, these ships perform sea control and limited power projection missions." Thanks, Bruce.

Iwo Jima's initial homeport was San Diego, California. In the summer of 1962 she took part in high-altitude nuclear tests at Johnston Island – the super-secret U.S. installation in the Pacific.

Following that assignment, Iwo Jima became part of a 21-ship squadron moved through the Panama Canal into the Caribbean during the Cuban missile crisis.

In September of 1963 she made her first of six WestPac deployments. During each of these, Iwo Jima played an active role in the Vietnam conflict, taking part in over 30 amphibious landings there.

In April of 1970 came notification of her designation as the primary recovery ship for the Apollo 13 mission. More on that spectacular mission a bit later.

In May 1972, she left San Diego and headed for her new homeport of Norfolk, Virginia, and was soon on her way to the Mediterranean for the first time and returned to Norfolk by January 1973. For the next twenty years Iwo Jima would make twelve Mediterranean deployments, a northern Europe deployment and many amphibious exercises.

Recognition came in the form of the Arleigh Burke Trophy as the most improved ship in the Atlantic Fleet in 1973. In October of that year she could be found clearing mines from the Suez Canal.

On what became her fourth deployment to the Mediterranean, she helped evacuate civilians from Beirut, Lebanon, in 1976.

In 1982 she earned her second Arleigh Burke award and in 1983 was part of the U.S. contingent of the Multi-National Peacekeeping Force off the coast of Lebanon from May to December, providing, at one point, medical support to Marines wounded in the tragic bombing of the Marine barracks in Beirut.

A regular overhaul took place at the Norfolk Naval Shipyard in July of 1986.

In 1990 – Desert Shield and Desert Storm. Iwo Jima became the first amphibious assault landing ship to deploy to such a sensitive area and remained there throughout the conflict. On October 30, 1990, a high-pressure steam leak on the ship cost the lives of ten of its crew; but despite that tragedy, the ship stayed on station after repairs.

In 1992 with war raging in the Balkans, Iwo Jima was in the Adriatic off the coast of Yugoslavia providing search and rescue support for UN flights as part of Operation Provide Promise. She was also on hand to provide medical support to the USS Saratoga and a Turkish destroyer after a missile-firing accident.

(Navy photo)

Iwo Jima was decommissioned on July 14, 1993, stricken from the naval lists in September of 1995, and sold for scrap in August of 1996.

We mentioned the two Arleigh Burke awards, but Iwo Jima was also awarded the Admiral Flatley Award for aviation safety four times, received many Battle Efficiency Awards for excellence in operations, amphibious assault, communications, aviation maintenance, damage control, and surface-to-air missile and gunnery exercises.

A postscript. The name lives on in the presence of LHD-7...also an amphibious warship but considerably larger in length and displacement. The current Iwo Jima was built at Pascagoula, Mississippi, and commissioned at the Pensacola Naval Air Station on June 30, 2001. She calls Norfolk, Virginia, homeport.

THE FLIGHT OF APOLLO 13 WITH JIM LOVELL, JOHN SWIGERT AND FRED HAISE FROM APRIL 11-17, 1970

Houston...we have a problem.
One of the great understatements of our time, and according to NASA records, not precisely what Jack Swigert told Mission Control after the explosion and vibration rocked the Command Module they had named Odyssey. Odyssey, by the way, means "any extended wandering" according to Mr. Webster; and if any mission tended to wander extensively, although never out of a prescribed flight path, it was Apollo 13. Actually Swigert said, "Houston we've had a problem here." And indeed they had.

For starters, any superstitious person had to be edgy simply because the number of this mission was thirteen, and NASA was hide-bound to hit the lift-off at 13 minutes after the hour. At Mission Control in Houston it was 13:13 p.m. CST, but, of course, at the Cape it was an hour later. There were no reported black cat sightings.

Although the big bang was not to come until they were 200,000 miles from Earth, there were ominous events almost from the start. The crew all felt a little vibration just five minutes after lift-off, as the center stage of the S-II engine shut down two minutes early. That had a ripple effect, calling for the remaining four engines to burn 34 seconds longer than planned, and the S-IVB third stage nine seconds longer to put Apollo 13 into Earth orbit. These were happenings not unlike a lot of other out of the ordinary events common to previous missions. Like them, the "anomalies" were confronted and dealt with so no one in the Odyssey or in Mission Control was overly concerned.

As a matter of fact, the journey to the Moon quickly became routine, almost boring, and one of the Earth-bound astronauts in Houston actually said, "We're bored to tears down here." It turned out to be the last time anyone mentioned boredom

The event that gave birth to books and a major motion picture and brought the news media back into the mission big-time happened nearly 56 hours into the mission and a long, long way from home. Jim Lovell had just wound up a 49-minute TV broadcast showing the viewing world how comfortable they were out there and had bid Earth goodnight. Nine minutes later, oxygen tank No. 2 blew up causing the No. 1 tank to also fail. In an instant...the Command Module's normal supply of electricity, light, and water was lost, and that can make any trip seem longer.

That is when Jack Swigert told Houston they had a problem, although none of the crew had any idea what the problem really was until sometime later. All they knew was they were suddenly without the necessities of life, and it rapidly became time to start thinking about, and carrying out, some alternatives. As has so often been said, "Truth is stranger than fiction." This mission drove the point home as never before. Thanks to the film "Apollo 13" and the intense press coverage throughout the now perilous journey, most of us can recollect bits and pieces of the improvisational genius, in space and on the ground, that combined to bring Apollo 13 and its crew home safely. Throughout my own years in radio and television, I have seen ample evidence that a product as basic as "duct tape" seemed to hold the industry together much of the time. Now we were about to be treated to its practically life-saving properties. Add cardboard to that list.

Only minutes after the explosion, Jim Lovell glanced out one of the CM's windows and realized there was some sort of venting coming from the Service Module. It was escaping oxygen and its loss was clearly evident in the CM, but it would be days before they would actually see the damage caused by the explosion. Only then, nearing home after all that incredible improvisation, would they separate from the LM, pull away, and take pictures of the massive damage the LM had sustained. It must have looked like a train wreck.

But for now that picture would have to wait. Time to close the hatch between the CM and the LM. It was a knee-jerk reaction, and quite logical since they could not know then exactly what was going wrong and why. As things turned out, they would survive only because they were able to spend most of their time in the Lunar Module, which has its own supplies of the life necessities they had just lost. There were ample supplies of oxygen, including the two EVA backpacks now not needed for Moonwalks.

LM had its own batteries, but they had not been expected to support life for the length of time it takes to go to the Moon and back. Water was the biggest problem. It was estimated they would run out of water about 5 hours prior to Earth re-entry, so in all cases, ultra-conservation became the watchword.

Carbon dioxide build-up was also a worry. We all exhale it when we breathe, and this accumulation now had to be dealt with in novel new ways. Incredibly, canisters of lithium hydroxide normally used to remove carbon dioxide from the Command Module were not compatible with those in the Lunar Module. This is where some of the incredible "improv" came into being and left an anxious world very, very impressed.

Enter the wonderful world of plastic bags, cardboard, and the indispensable duct tape. Rube Goldberg would have been proud of the way Mission Control advised the crew to construct this homely devise and how beautifully it worked.

Suddenly a flight path originally designed with a Moon landing in mind would have taken dangerously long, considering all the strikes that Apollo 13 now had stacked against it. A never-before tried burn of the Lunar Module engine had to be chanced to speed up their return path home by at least ten hours. Truly, the LM had become Apollo 13's lifeboat in ways never imagined. That LM decent engine, and not the SPS engine in the Service Module, did this badly-needed step nicely after rocket specialists at TRW Systems, its builders, did some fast test-firing and reported back that it could do the job with plenty to spare.

Imagine the activity, the thinking, and the totally new directions now demanded of everyone – not just Mission Control, but among the many companies like TRW, which designed and built the countless components that go into a lunar flight. Suddenly, there were teams of experts working out course corrections and new detailed timelines. The Lunar Module was never designed to stay attached to the Command and Service Module nearly all the way back to Earth, but now it would, and its propulsion systems used in ways never dreamed of. It would be

jettisoned (and then, pictures taken of it) only a few hours before re-entry.

In the meantime, life in the LM was approaching the primitive. With all systems powered down to ensure their longevity, the CM and the Lunar Module were just plain cold. Condensation on the walls gave rise to fears of an electrical short-circuit. The men of Odyssey were tired and cold, and the film "Apollo 13" would later portray that situation quite realistically. Despite their fatigue-induced mistakes, magnified by the inability to take accurate star sightings (due to the gas cloud escaping from the Service Module), the crew was able to use Sun alignment to verify their re-entry path. When the time came, the last burn before re-entry was…perfect.

This mission just kept piling up its small miracles and nothing dramatized the wonder of Apollo 13 more than the splashdown – just four miles from the USS Iwo Jima, one of the closest ever to a recovery ship. Captain L. E. Kirkemo commanded Iwo Jima. Helicopter Squadron 4 was on hand for its fifth and last mission, and piloting the Recovery helo was Commander Chuck Smiley. The frogmen from UDT-11 were ably led by Lt. Ernie Jahncke. Recovery took place in the Pacific, of course, southeast of Samoa.

No sooner had man landed on the Moon and then did it again on the very next attempt, public interest inevitably began to fade. One sign of that came when the networks went back to their standard two-reporter coverage right after the Apollo 11 flight when we were a foursome.

Now Apollo 13 had the big audiences back again, but barring any further hair-raising adventures, the surveys would quickly show complacency in the listening, viewing, and even the reading world. Safe airline departures and arrivals don't make news. Airline crashes do. It has been ever thus.

In the meantime, Odyssey itself did not seem any the worse for its ordeal. You can find it sitting comfortably at the Kansas Cosmosphere and Space Center in Hutchinson, Kansas, about 30 to 40 minutes north of Wichita.

John Swigert and Fred Haise were civilians during this amazing journey.

USS NEW ORLEANS (LPH-11)

The USS New Orleans (LPH-11) became the primary recovery ship for not just Apollo 14 but for the last three splashdowns of the manned spacecraft program – Skylab 3 and 4 and Apollo-Soyuz.

The LPH class of carriers came into their own as the manned recovery phase of the space program wound down. Of the last nine splashdowns only the USS Ticonderoga was not an LPH. New Orleans matched her measurements with Iwo Jima.

A BRIEF HISTORY OF THE USS NEW ORLEANS (LPH-11)

The USS New Orleans (LPH-11) was the fourth ship to bear the name but the first under that name to be built as a carrier. For the record, the first New Orleans emerged during the War of 1812, (there was a Confederate New Orleans in the early 1860's) a cruiser built in England but purchased by the U.S. Navy in 1889 and a heavy cruiser built in New York City in 1931 which survived the Japanese attack on Pearl Harbor and had an incredible war record…ending the war with 16 battle stars.

This USS New Orleans barely got into the Vietnam War, having begun life when her keel was laid down at the Philadelphia Naval Shipyard on March 1, 1966. Launching occurred on February 3, 1968, and commissioning on November 16, 1968.

She was to become the second most active carrier in the U.S. manned spacecraft program right behind the USS Wasp. One Apollo…. two Skylabs and one Apollo-Soyuz splashdown. She literally served out that part of the U.S. space program, as you will read.

Her first appearance was in the western Pacific in August of 1969 as flagship for the Amphibious Ready Group Bravo. In October she hosted the Eighth Vietnamese awards and later that month took part in Operation Keystone Cardinal, a retrograde movement of Marines out of Vietnam, so she was on the "outer" edge of that fading conflict. New Orleans headed back to San Diego in the spring of 1970.

By August she was flagship for Commander First Fleet, western Pacific and provided support and security for President Nixon's visit to Mexico. Late that year she began preparing for her first NASA involvement; and on February 9, 1971, she was there to pick up the astronauts of Apollo 14, Alan Shepard, Stuart Roosa and Ed Mitchell some 900 miles south of American Samoa.

New Orleans made her next WestPac deployment and went through a series of training exercises including a multi-

(Navy photo)

national cruising exercise simulating a convoy under combat conditions as well as using her helicopters in a simulated assault with the Marines on an island in the Philippines.

Into 1973 and New Orleans was tasked with de-mining operations for the coasts and harbors of North Vietnam.... an operation which ended in mid-April.

The Chief of Naval Operations had lengthy plans in mind for her by July of that year, as she became the last carrier to effect the recoveries of our last three space missions prior to the start of the Space Shuttle program.

Back to regular assignments in 1980 with several months spent in the Indian Ocean during the Iran hostage crisis. Big overhaul at Bremerton, Washington, in 1981 and various exercises in the Indian Ocean and western Pacific extending to 1986. She finished out the 80's with a number of amphibious exercises throughout the western Pacific, marking her 13th deployment into that theater.

From late December of 1990 to late August of 1991, the New Orleans could be found in the Indian Ocean and Persian Gulf in support of Operations Desert Storm and Desert Shield. In that deployment she served as part of the largest amphibious task force to sail from the west coast of the U.S. in 25 years. On the first day of ground fighting in Iraq she off-loaded 1700 Marine combat troops and then tackled aviation mine countermeasures in waters ten miles off the coast of Kuwait.

November 1991, and the USS New Orleans was in her 23rd year. It was time for a really big overhaul. Six months in San Diego and then out for, among other tasks, her 15th western Pacific deployment and a part in operations off Mogadishu, Somalia, in 1993.

In November 1994, Hollywood came calling and her decks welcomed film director Ron Howard and actors Tom Hanks, Kevin Bacon and Bill Paxton for the filming of several portions of the Oscar-winning movie "Apollo 13."

The clock was running for the New Orleans; and as thirty years of superb service neared, she sailed for the western Pacific one more time and put a cap on it in May of 1997. She was decommissioned and placed in reserve in San Diego in November of that year. Look at her record at retirement. The Navy Unit Commendation, four Battle Efficiency Awards, the Meritorious Unit Commendation, the Navy Expeditionary Medal, the Armed Forces Expeditionary Medal, the Southwest Asia Campaign Medal, the Vietnam Service Medal and the Kuwait Liberation Medal. All that and four splashdowns. Good show.

THE FLIGHT OF APOLLO 14 WITH ASTRONAUTS ALAN SHEPARD, ED MITCHELL AND STUART ROOSA FROM JANUARY 31 TO FEBRUARY 9, 1971

More than eight months after Apollo 13 had us all on the edges of our seats, three more American astronauts couldn't wait to get off the launch pad and try it again. In fact, NASA even decided to let Apollo 14 land in the lunar region which Apollo 13 never had the opportunity to do. Fra Mauro.

You didn't have to be in the space program in those days to figure out that each lunar landing would probably involve longer and more distant walks, or rides, on the Moon. Well, the men of Apollo 14 wouldn't be able to drive around up there yet, but one of their number did drive a golf ball...and they did bring along a sort of rickshaw to haul their scientific gear around. Mankind was still, "taking giant steps each time out."

There was added interest in the mission since America's first man in space, Alan Shepard, had been grounded by a medical problem after his historic sub-orbital flight. He was visible and extremely popular. His ability to overcome that setback paved the way for him to realize his dream; and being Alan Shepard, he had devised a way to put an asterisk beside his name before it was over.

Apollo 14... lift-off at 4:03 p.m. from Launch Pad 39A on the last day of January 1971, after a brief delay to let some rain clouds pass through the Cape area. It was, by the way, the first such delay in the Apollo program. This time the Command Module answered to the name Kitty Hawk, while the Lunar Module stowed below the CM was known as Antares, which is the brightest star in the Scorpio constellation. But you knew that.

What would one of these journeys be without a problem here and there? Exactly. Therefore one might have anticipated the flight to the Moon would not be without a hitch or two. During transposition and docking, that is, opening up the shroud covering the LM down in the Service Module, bringing that spidery guy out of there, turning the CM around and locking the two together, they certainly did have problems. It took six tries before the two became a docked and locked unit—nearly two hours longer than planned. Of course, they had to take a long look at the point where the two spacecraft came together, but they could not determine why it took so long to do the job. Nor did they see anything that made them think it wouldn't work just fine later on.

It was shortly before 2 a.m. on February 4 when the Kitty Hawk/Antares combination entered lunar orbit. About 45 minutes later, the crew kicked out their SIV-B stage and sent it crashing into the Moon where the Apollo 12 seismometer, sitting there since November of 1969, duly reported vibrations from the crash for over two hours. That was some of that ALSEP package from Apollo 12 doing its job. This mission carried much of the same equipment. Soon after a good lunar landing at 4:18 a.m. on February 5, Alan Shepard and Ed Mitchell were on the Moon's surface spreading the ALSEP gear all around them; namely, a TV, an S-band antenna, our flag, a Solar Wind Composition experiment and more.

Using cameras most photo fans would mortgage the farm for, they took pictures of themselves going about their lunar chores. Shepard's surprise was not to be unwrapped until EVA number two.

On this initial Moonwalk, they unloaded and set up the rickshaw we spoke of earlier. A pull-cart enabled them to retrieve some Moon rocks they might not have been able to bring back in bags. Shepard actually grabbed a couple of rocks the size of footballs, and this time there was a convenient place to put them. They ventured out some 500 feet from the LM to set up many of their experiments. Four and a half fun-filled hours later they were back in Antares and

ready for sleep…or as close to it as they could get. At least they had hammocks.

Only Moon walkers could ever understand some of the difficulties Shepard and Mitchell faced on their second EVA. Depending on the sight lines and the sun, there were times when they could see craters in one direction and not in another – even when those craters were very close. At any rate, it was another walk around of more than four hours, almost a hundred pounds of rocks and hundreds of photographs. Not a bad day's work. However, Alan Shepard was not done yet. Reaching into a zippered pocket, he produced a little white object familiar to many of us…a golf ball. He dropped it on the dusty lunar surface, and using the handle of the contingency sample return container with what he swore was a "genuine six-iron head" attached, he took a couple of one-handed swings. Missing the first and making all us duffers feel equal, he proceeded to hit his second try for what was by all possible measures, a very handsome shot. He then reported that it went for "miles and miles" and, for all we know, it might have. You can rest assured that none of us have seen a six-iron shot go as far down here.

Of course, it was unauthorized, as was a famous corned beef sandwich years before, but in both cases, viewers just ate it up; and it became the talk of the town, city, and the country. Even NASA officials, who were aghast at the sandwich situation where the crumbs could have really created problems, had to roar and high-five in Mission Control when they saw what Shepard had done. Too bad he could not have gone out and retrieved that shot. A sure bet for placement in Golf's Hall of Fame, but Alan's six iron IS there. The golf shot heard round two worlds came at the very end of their last Moon walk, so he respectfully unveiled it at a point where it did not become a major distraction.

Meanwhile, back in Kitty Hawk, our intrepid CM pilot Stu Roosa had his camera out much of the time and was clicking away at prospective future landing sites, such as Descartes, and writing down prominent lunar landmarks that would improve landing accuracy on later missions.

Now it was time to get back together. On the lunar surface Shepard and Mitchell packed away their samples and threw out expendable equipment which, in come cases included a Hasselblad or two. I took a picture of one that did make it back while I was aboard the USS Hornet, just after the return of Apollo 11. (See page 76.) A NASA photographer said, "Try 5 grand for that baby." Worth its weight in Moon rocks? NASA thought so.

Lunar lift off … 1:48 p.m. February 6. There's a tendency to picture the LM, minus its descent engine now, merely rising straight up for 50 or 60 miles and docking with the CM. Actually it involved gradually merging as their paths narrowed and the chase across the lunar surface took fully two and a half hours before LM and CM were back together. Thirty minutes later, Antares was given its walking papers and sent crashing back to the Moon. Sounds absolutely ungrateful, doesn't it?

On the way home, the crew carried out a series of interesting experiments, which were promising enough to be slated for future flights on Skylab and the Space Shuttle. Other than that, just another routine trip to the Moon and back. Re-entry was primarily computer-guided and normal. Splashdown in the South Pacific was less than a mile from their target point and comfortably close to the recovery carrier USS New Orleans, which would become the Navy's workhorse during Skylab. The skipper was Captain R. E. Moore. The Helicopter Squadron was HS-6 and the prime recovery pilot was Commander W. E. Walker. UDT-11 sent out its frogmen for the fourth time. With time out of the abortive Apollo 13 missions, those Mobile Quarantine Facility trailers (MQF's) aboard the carrier came back into play, and the three astronauts and two NASA technicians would spend three days inside it before entering the Lunar Receiving Laboratory in Houston. That would do it. It was the last mission in which a crew would be quarantined. NASA had become satisfied that the Moon harbored us no ill will or germs unknown to mankind.

If you would like to feast your eyes on Apollo 14, Kitty Hawk, and are planning a trip to Disney World—or just Florida any time soon – include the Cape in those plans and check out the Astronaut Hall of Fame in Titusville, just north of the Cape itself. The Apollo 14 MQF is on public display aboard the USS Hornet Museum in Alameda, California.

Stuart Roosa was an Air Force Colonel and Ed Mitchell was a Navy Captain when they flew on Apollo 14.

USS OKINAWA (LPH-3)

The USS Okinawa became the prime recovery ship for the Apollo 15 mission with astronauts Dave Scott, Jim Irwin and Al Worden from July 26 to August 7, 1971.

Okinawa was yet another LPH, an amphibious assault ship designed to get combat-ready Marines to trouble spots in a hurry. Her numbers were basically the same as those of her sister LPH's.

A BRIEF HISTORY OF THE USS OKINAWA (LPH-3)

Okinawa began life when shipbuilders at the Philadelphia Naval Shipyard laid down her keel on April 1, 1960, and launched August

(Navy photo)

19, 1961. She was commissioned on April 14, 1962, and would provide more than thirty years of fine service to the Navy.

Following sea trials, Okinawa headed for her homeport of Norfolk, Virginia, in June of 1962, and by that fall was placed in service as part of the fleet enforcing the naval quarantine during the Cuban missile crisis.

Over the next several years, Okinawa had several deployments in the Caribbean but found time for a good will visit to the New York World's Fair in the summer of 1964. Later that year she participated in an amphibious exercise, or maneuver, off the coast of Spain – Steel Pike I.

In April of 1965 she was called upon to stand by as a medical evacuation ship during a domestic crisis in the Dominican Republic. After that it was back to Norfolk for overhaul and further deployment in the Caribbean.

Vietnam was in the headlines and Okinawa was transferred to the Pacific fleet in early 1967, arriving in her new homeport of San Diego on February 8. On March 10 the carrier left for her first deployment off Vietnam; but while sailing from Okinawa to Taiwan on April 13, she was diverted by a distress call from a grounded Panamanian freighter, the Silver Peak. The next day Okinawa successfully rescued all 38 persons on board that ship.

She returned to Vietnam duty for the next several months but returned to San Diego in early December. She was about to get her feet wet in the space program.

On April 4, 1968, after an intensive period of special training, Okinawa recovered the unmanned

Apollo 6 space capsule in the Pacific some 380 miles north of Kauai, Hawaii.

By November she was headed back to the western Pacific and during the next six months, into the middle of 1969, became heavily involved with her Marine forces in search and destroy operations on Viet Cong rest camps on an island a few miles south of Danang. Then, back to San Diego in late June for leave and maintenance.

During her Vietnam tenure, Okinawa led six amphibious assaults and helped evacuate Phnom Penh and Saigon while rescuing downed American pilots, one of whom, Randy "Duke" Cunningham, went on to represent San Diego in Congress for several terms.

A 7th Fleet cruise in 1970 took her to the Philippines, New Zealand, Okinawa, Japan and Vietnam.

In mid-1971, Okinawa was chosen to recover the crew, Command Module, and lunar samples of Apollo 15.

It was April of 1975 when the North Vietnamese overran much of South Vietnam, and Okinawa was but one of many ships called on to get U.S. and other allied forces out of there. South Vietnam officially surrendered on April 30.

Late 1970's…. a WestPac cruise.

In mid-1980, Okinawa was sent to the northern Arabian gulf to support Afghanistan contingency operations, another WestPac deployment in 1981 before heading back once again to San Diego.

1990-91 – Activity in support of Operation Desert Shield and Desert Storm where her Marine forces liberated an Iraqi-held island off the coast of Kuwait.

1992 saw Okinawa back in the western Pacific for her last deployment. She was decommissioned on November 19, 1992, at the San Diego Naval station after 30 years of service, stricken from the Naval Register on December 17 but retained as what was called a "parts hulk."

I was an Army guy in the mid-50's but I nonetheless feel saddened and wonder what the men who actually served and sailed on these great ships feel when they learn that "their" ship has been relegated to target practice. Okinawa was sunk as a target on June 6, 2002, off the coast of Southern California. She rests in 2,000 fathoms of water.

She did not go quietly, nor did she serve those three decades without notice and recognition. Her many awards included the Armed Forces Expeditionary Medal, the Combat Action Ribbon, the Armed Forces Service Medal, the Navy Expeditionary Medal, the Humanitarian Service Medal, the Meritorious Unit Commendation, Navy "E" Ribbon, Navy Unit Commendation and the Southwest Asia Service Medal, the Vietnam Service Medal, the Republic of Vietnam Meritorious Unit Citation, and the Republic of the Philippines Presidential Unit Citation.

Okinawa sleeps well.

THE FLIGHT OF APOLLO 15 WITH ASTRONAUTS DAVE SCOTT, JIM IRWIN AND AL WORDEN FROM JULY 26 TO AUGUST 7, 1971

Rover arrives on the Moon! After several years in the design and construction (by Boeing), the Lunar Rover made its debut on this mission, and was worth every dollar that NASA had to spend. When the tested and accepted design went into "production" just three were built at a total cost of around 40 million. No, they were not "on special" that year.

Apollo 15 continued NASA's planned policy of expanding the scope of every mission to reach a little farther in actual distance and in scientific knowledge gained. Consequently, the space agency was ready to put its lunar pair on wheels to go out miles instead of feet and yards and still be able to get back to the LM, named Falcon for this mission. The orbiting Command Module cruising overhead was called Endeavor, a name that would live on into the Shuttle program a decade later.

When I lecture on the space program, I use, among dozens and dozens of slides I took myself or obtained from NASA, several of Dave Scott using a battery-powered drill on the Moon – the Black and Decker creation. It was only when I dove into writing this book that I came to realize the drilling part of the mission was no piece of cake. But, like all other aspects of our space program, Scott's difficulties paved the way for modifications and improvements enjoyed by both Apollo 16 and 17, the last manned lunar missions. More on that coming up a bit later.

Getting back to Earth and the Apollo 15 lift-off. It was 9:34 on the morning of July 26, 1971, on what would be a lunar mission approaching 13 days in length, when Apollo 15 headed for the Moon from Launch Complex 39A with no unusual delays or problems enroute.

While the entire complex from launch vehicle to Command Module might have appeared exactly the same as all its predecessors, there were unseen improvements which accommodated the Boeing Lunar Rover tucked underneath the Falcon lunar lander (LM) inside the Service Module. This versatile little vehicle had to have extraordinary capabilities, weigh as little as possible, and be controllable by astronauts whose own mobility was severely restricted by the ever-bulky spacesuits.

At one point late in its development, Boeing had given the Rover a pistol-grip single control between its two seats. Astronauts, who expected to drive it someday, were quick to point out that with their bulky, rigid gloves, the pistol grip would prove impossible to work with. The pistol grip gave way to a simpler T-shaped hand controller that gave a driver steering, speed and braking all at one point. It had to be that way. Because its simplicity made it a one-hand operation, the system would one day migrate into Earth-bound vehicles designed to be driven by the severely handicapped…amputees, paraplegics and the like. Just one of thousands of products now making all our lives better but originally developed for the needs of space flight.

Getting back to the business of lunar exploration. The usual process of unpacking the LM and getting it around to nose-to-nose docking with the Command and Service Module took place during the several days it took to get to the Moon. Lunar orbit insertion was uneventful and Falcon, with astronauts Dave Scott and Jim Irwin aboard, touched down in the Moon's Hadley Appenine region at 6:16 p.m. on July 30.

At that time, Scott and Irwin had been up for 11 hours and any thoughts of going right out to the lunar surface were put to bed with the astronauts. While sleep was as fitful as it had been on all previous flights, it was a smart move. Before turning in however, they did give Mission Control a thorough description of their new surroundings. They put on their helmets and gloves, opened the hatch and did a stand-up EVA. Dave Scott stood on the cover of the ascent engine, head and arms outside the spacecraft, and made good use of his 70 mm Hasselblad with a nifty 500 mm lens.

Then another NASA first took place in the Falcon Lunar Module. The astronauts actually got out of their bulky spacesuits and got really comfortable in those hammocks. It had to be the best sleep yet for any of our Moon explorers.

They had an ALSEP scientific package on board, but their first job was to deploy the Rover…. get it out from under the spiny legs of the LM and ready to roll all over the Moon. This multi-million dollar hot-rod had wire wheels. Nothing special you say? No, really…. wire-mesh wheels. No rubber tires here. Wire-mesh to sink into and grip the ever-changing gritty soil of the Moon. It had four-wheel, all-electric drive with a quarter horsepower motor on each wheel. Think about it. A gasoline engine needs air to breathe and function. No air up there. As it sat, it weighed in at 460 pounds, but with two passengers and a lot of equipment, it came in at more like 1500 pounds. With all that it could tear along at a zippy 6 to 8 miles an hour, but that was on a smooth surface. There was nothing smooth or predictable about the Moon. It also had an absolutely necessary navigation system to tell the two astronauts where they were and, at all times, how far they were from home…. the lunar lander. On the Moon, it was oftentimes difficult, if not impossible, to gauge their distance from Falcon. Many times there were rises and craters which completely obscured their view of that home base. Even with a direct sight line, it was never easy to estimate their distance from it.

In the area they were in, or could now drive to, there were mountains as high as 11 to 14 thousand feet, and some had slopes of around 30 degrees, which would have made them challenging ski slopes here on Earth. Scott and Irwin visited them during their three

EVA's and collected rock and soil samples, which, at least on initial examination, seemed quite different from those already gathered near the LM. At one point they found a rock that would later be named the "Genesis Rock" because scientists calculated its age as somewhere in the 4 billion years range. It was a crystalline rock made up almost entirely of a mineral called plagioclase.... unique. As astronaut Jim Irwin commented to Mission Control in Houston, "I think we found what we came for."

They had hoped to do better with the newly developed battery-powered drill they brought with them, but each time Dave Scott tried to use it in hopes of getting down into the lunar surface 6 or 7 feet and bring back core samples, the drill would bog down and virtually stall at no more than, say, four or five feet. Instead of spiraling the sub-surface material back up through it and to the surface for collecting, the tubing or flutes simply jammed and nothing could move. It was a heavy duty and frustrating effort, but once again, the experience of this mission would lead to far better performances on the last two lunar missions.

Despite their disappointment, Scott later found his efforts had penetrated over fifty distinct layers, and the scientists to whom these experiments were turned over were nothing less than elated. It would only get better.

Late in the mission they tried, unsuccessfully, to do a Rover drive-around with Dave Scott at the wheel and Jim Irwin following him with the movie camera in what was to be a show for Boeing and Marshall engineers. The movie camera picked that moment to not work. It was just an added assignment for Apollo 16. Everyone already knew the Rover had delivered on its promise anyway.

So there were stops and starts, triumphs and disappointments, but, all in all, Scott and Irwin had accomplished much and had not exceeded any practical limits to the length of an Apollo mission. Lessons were learned, as always, and improvements made.

With the break-through use of an electric-powered vehicle on the Moon, the amount of rock and soil samples was growing with every mission. From Apollo 11's 46 pounds to 75 on Apollo 12, 94 on Apollo 14 and now nearly 170 pounds. The heaviest loads were yet to come.

Waiting patiently in the South Pacific was another well-rehearsed and ready recovery ship, the USS Okinawa, the latest in the line of Iwo Jima class amphibious assault ships. The skipper was Captain Andy Huff. I was on that ship and it would turn out to be my last assignment as a "pool" reporter. Apollo 15 would descend on those highly visible three orange-and-white parachutes right in front of us but about 5 or 6 miles out. Still, it was close enough and clear enough that, even as I was broadcasting the descent of the CM, I was clicking off a few color photos of the event at the same time. I actually photographed the moment of splashdown, but at that distance, the finished slide did not show it clearly. At splashdown we were just about 330 miles north of Honolulu.

Helicopter Squadron HC-1 was just beginning what would become a string of six straight recoveries, including all three Skylab missions. Commander Steve Coakley led the squadron. UDT-11 was back to provide the frogmen, led by Lt. Fred Schmidt.

No more MQF, no more bothersome quarantine. The fit and fine looking trio emerged from one of the recovery helicopters onto Okinawa's deck an hour or so after that splashdown and walked steadily down a red carpet to mid-deck and a battery of microphones where they each took turns thanking the Navy and NASA for putting the final touch on a fine mission. They later joined us in the officer's mess for one more excellent meal on what was the finest food ship I had sailed on in my five recoveries. They were all pretty good...Okinawa was excellent.

There was one more rather special moment left in this mission when a group of us noticed something floating just beneath the surface as the Command Module was being pulled by cable to one of Okinawa's aircraft elevators. We shouted to the young men in a ship's dory to grab it, whatever it was, and save it from going to the bottom of the Pacific. It turned out to be the CM's protective heat shield, which we in the news media persisted in calling a nose cone. Some folks at the National Air & Space Museum didn't know what I meant by "nose cone," but they burst my balloon of reality when they assured me this retrieval was not the first in splashdown history. Nonetheless, we got it back, and I'm still proud of the pictures I have of it.

Apollo 15 now calls the U.S. Air Force Museum at Wright-Patterson Air Force Base in Dayton, Ohio, – home.

Jim Irwin and Al Worden were both Air Force Colonels.

USS TICONDEROGA (CVS-14)

Ticonderoga became the prime recovery ship for the last two Apollo Moon missions – Apollo 16 and 17 and the first Skylab, known as Skylab 2.

Ticonderoga. The name says it all. As a pacesetter built with certain improvements over the classic Essex carrier design, her name became the "class" designation of all carriers that underwent the same basic changes.

A BRIEF HISTORY OF THE USS TICONDEROGA (CVS-14)

Having read this far you know that Ticonderoga class meant a length of 888 feet and a bow specially modified to accommodate those badly needed anti-craft guns. She would come to share another commonality with many of her sister carriers. When her keel was laid down on February 1, 1943, she was the Hancock; but she had already become the first Ticonderoga class carrier by the time she was reassigned that name on May 1. She was launched on February 7, 1944, and commissioned at the Norfolk Navy Yard on May 8 of that year.

Ticonderoga did her rehearsing for war primarily in the Caribbean over the summer of 1944 but headed for the Pacific in the fall and had become part of famed Task Force 38 by late October.

She quickly became a formidable presence in the effort to re-take the Philippines. Her toll of Japanese planes both in the air and on the ground grew rapidly. In the ensuing months Task Force 38 attacked and sank dozens of Japanese ships – troop and supply transports, and warships from destroyers to heavy cruisers.

Typhoons raged through the area several times late that year and one of them cost the Navy three destroyers and 800 men. The storms actually forced TF 38 into anchorages for repairs and the group did not return to sea until December 30.

(Navy photo)

The power of that Task Force was seen on January 12, 1945, when its carriers, Ticonderoga in their midst, launched not less than 850 planes which proceeded to sink a whopping 44 ships. The Japanese Navy was staggering.

Ticonderoga was badly damaged by kamikaze planes on January 20 and 21, losing several hundred sailors, the ship's Captain Kiefer, and many aircraft. She limped into the seldom talked about anchorage at Ulithi but only long enough to remove her wounded before heading for major repair work at the Puget Sound Navy Yard in Washington state. She was there from February 15 to April 20. Back in the Pacific she rejoined the Fast Carrier Task Force as an element of Task Group 58.4.

Now the targets for her planes were both the Japanese home islands or nearby islands such as Okinawa, where American ground forces were locked in intense fighting.

These attacks continued right on into August when the atomic bombs dropped on Hiroshima and Nagasaki on August 6 and 9 respectively, hastened Japan's capitulation a week later.

Like her sister ships, the crew of the Ticonderoga had to transition from the urgency of wartime operations to the proceduralism of peacetime missions. She continued her patrols in Japanese waters while her planes went searching for camps still containing Allied prisoners of war. Supplies were airdropped.

What followed was that series of transport voyages dubbed "Magic Carpet" and it must have been magic for the thousands who found themselves heading home aboard Ticonderoga and her sister carriers over the remaining months of 1945. In January 1946 she entered the Puget Sound Naval Shipyard to prepare for inactivation and almost a year later was placed out of commission and berthed with the Bremerton Group of the Pacific Reserve Fleet.

Her story certainly did not end there. Out of reserve on January 31, 1952, Ticonderoga was decommissioned soon afterwards to undergo extensive conversion at the New York Naval Shipyard. Extensive? How about 29 months? Steam catapults to launch jets, a new deck-edge elevator and whatever else was new and available in electronics.

January 1955 saw Ticonderoga shifting to her new home port of Norfolk, Virginia, but was soon back in New York for even more alterations followed by tests for three new planes, eight months in the Mediterranean and back to Norfolk on August 2, 1956, for the angled flight deck and an enclosed hurricane bow.

Over the next eight years, Ticonderoga would be in and out of the western Pacific conducting not only training operations with the 7th Fleet but making many goodwill and liberty port calls throughout the Far East.

On August 2, 1964, she got her first taste of Vietnam when her aircraft attacked units of the North Vietnamese Navy in international waters in the Gulf of Tonkin. This and other clashes just days later led President Lyndon Johnson to respond with attacks on selected North Vietnamese torpedo boat bases, and the long, controversial Vietnam war was on. For her quick reaction and successful combat actions on those three occasions, Ticonderoga received the Navy Unit Commendation.

There came a time-out when Ticonderoga returned to the U.S. and the Hunter's Point Naval Shipyard in California in late January of 1965 for a five-month overhaul. Working her way back to the Pacific, she was back on Dixie Station (South Vietnam) on November 5 and combat air operations began immediately.

It becomes easier to understand why and how Ticonderoga earned 12 battle stars during the Vietnam conflict, since the ship was destined to make six combat deployments to that area by June 1969. The combat sorties flown by her planes soared into the many thousands, as did the tons of ordnance dropped on North Vietnamese targets. There were several returns to the states for repairs and replenishments, but it was always…. back to Vietnamese waters.

Ticonderoga arrived back in San Diego on September 18, moving to the Long Beach Naval Shipyard in mid-October to begin conversion to an ASW, anti-submarine warfare carrier and hence, the latest and last designation - CVS-14.

This was not to be the end of her deployments to the Far East however. Her change of mission meant that her two deployments into that theater no longer included combat operations off Vietnam. In between those last two deployments, Ticonderoga recovered the crew of Apollo 16 in April 1972, near American Samoa. More training exercises and then on December 19 the recovery of the Apollo 17 crew – the last Moon landing mission.

Back to San Diego on December 28 and nine more months of active duty, which came to include picking up the astronauts of the first manned Skylab mission, Skylab 2 on May 25, 1973.

On September 1 the carrier was decommissioned after a board of inspection and survey found her to unfit for further naval service. Her name was struck from the Navy list on November 16 and arrangements were begun to sell her for scrap.

The bottom line? Five battle stars during World War II, three Navy Unit Commendations, one Meritorious Unit Commendation and 12 battle stars during the Vietnam War.

THE FLIGHT OF APOLLO 16 WITH ASTRONAUTS JOHN YOUNG, TOM MATTINGLY AND CHARLES DUKE FROM APRIL 16 TO THE 27, 1972

At one point in the space program, NASA was prepared to go to an Apollo 20, but, as public interest inevitably waned as it always does when the news is good and tragedy does not intervene, the space agency was coming under increasing pressure to call it a day on the lunar excursions. Don't you guys have enough Moon rocks already?

Yes, the public interest curve was heading decidedly downwards almost as though going to the Moon was like a walk in the park. Been there, seen that. The assignment list had taken on a carbon copy look of missions before. Go out, collect soil and rock samples, set up the ALSEP scientific experiments, take pictures and drive that snappy little lunar Rover. The single factor, which separated these missions from each other, was the fact that NASA sent the astronauts to different parts of the Moon each time.

For Apollo 16, it would be the Descartes region, which a Command Module pilot, (Stuart Roosa on Apollo 14) waiting for his colleagues to return from the lunar surface, had extensively photographed over two years earlier. John Young who was in the midst of a remarkable space-flight career would command this mission. Young had already flown two Gemini missions. Young was also on the Apollo 10 lunar orbit mission and would go on to pilot two Shuttle missions including the first in 1981. His teammates on Apollo 16 were both rookies.

I doubt that any of the three men were overly concerned with the decline in public and congressional interest in Moon missions when, on the 16th of April, 1972, at just a few minutes before one o'clock in the afternoon, they lifted off from Launch Pad 39A in the Command Module known as Casper. The Lunar Module was named Orion.

A few, not insurmountable problems occurred on their way to the Moon but had been solved and put behind them long before they entered lunar orbit on April 19. Next day John Young and Charles Duke entered the LM and separated from Tom Mattingly in the Command and Service Module but the actual descent to the Moon had to be delayed 6 hours because of technical problems. Engineers looked at it and decided it would not seriously affect CSM steering, which had been the threat, so the mission proceeded and the LM headed down to the lunar surface at 2:23 U.T. on April 21. Now, you ask, what is U.T.? It pops up in NASA's mission histories now and then and it stands for Universal Time. These days it is often referred to as GMT or Greenwich Mean Time. In any case, it is five hours ahead of Eastern Standard Time so that 2:23 up there was, in our reality, approaching 9:30 p.m. the night before, or April 20. Are you still with us?

Interesting. As the poundage in lunar soil and rocks piled up throughout these Apollo missions, so did their time spent out on the lunar surface and spinning around in the Lunar Rover. Apollo 15's three EVA's added up to 18 hours. Now, Apollo 16 was about to give us more than 20 hours and collect over 210 pounds of rocks and stuff. No sign of green cheese as yet.

In their three EVA's and Lunar Rover excursions, Young and Duke put over 15 miles on their little vehicle. Do you think if the Rover had been ready for the Apollo 14 mission, that lunar golfer Alan Shepard really would have jumped in and gone out to get that golf ball? World's first out-of-this-world golf cart. You can bet Shepard thought about that as he took part, from Mission Control, in the missions following his flight.

Young and Duke deployed all the devices in the ALSEP package, collected lots of rocks as noted above and, of course, took lots and lots of pictures. Meanwhile, their CM pilot, Tom Mattingly, had his own marching orders for other in-flight experiments and photography from his lunar orbit.

The Lunar Module Orion lifted off from the Moon at 01:25 U.T. (now you know what it is, you have to figure it out) on April 24 after more than 71 hours on the lunar surface. LM and CSM were back together some two hours and ten minutes later. Then something strange happened to the mission. LM was jettisoned according to plan about 15 hours later, but instead of crashing back into the Moon and giving those left-behind seismometers something to shake about, an open circuit breaker in the guidance and navigation system prevented that from happening. Therefore, the LM remained in lunar orbit with an estimated lifetime of one year. No, the next, and last, mission to the Moon did not see it again.

All went well for the trip home. Command Module separation from the Service Module occurred on April 27 at 19:16 UT. There's that strange time again. Would you believe 2:16 EST? Now that Earth was coming back into the story, NASA went back to time references we Earthlings could relate to. Splashdown in the Pacific, 215 miles from Christmas Island came at 2:45 p.m. EST on the 27th. They had been in space for more than 265 hours. They landed (splashed) only 3 miles from the USS Ticonderoga, which would remain the primary recovery ship for the next two missions. Ticonderoga's captain was E. A. Boyd. Helicopter squadron HC-1 was back in action led by Commander Arnie Feiser piloting Recovery 1. The frogmen belonged to UDT-12 led by Earl Kishida.

Apollo 16 found a home at the U.S. Space & Rocket Center in Huntsville, Alabama.

At the time of this mission, Tom Mattingly was a Navy Captain. Charles Duke was an Air Force Colonel.

THE FLIGHT OF APOLLO 17 WITH ASTRONAUTS GENE CERNAN, RON EVANS AND HARRISON SCHMITT FROM DECEMBER 7 TO 19, 1972

Last manned mission to the Moon – USS Ticonderoga was again the recovery ship.

Despite the dwindling interest on the parts of the public and the press, I would have gladly jumped aboard the USS Ticonderoga just to have been there to see the last astronauts to go to the Moon and return safely home. Gene Cernan, who would go into the history books as the last man to walk on the Moon, had been one of the first astronauts I watched return from Earth orbit when I covered my first splashdown reporting assignment aboard the USS Wasp for the Gemini 9 mission in June of 1966.

In 1994 I was given the honor of emceeing a black-tie affair in Manhattan on the occasion of Apollo 11's 25th anniversary and Gene Cernan sat next to my wife and I at one of the head tables. He was congenial, entertaining and classy and he and I engaged in a bit of repartee that none of us at that table would ever forget. More on that later.

NASA must have had someone in authority who liked to tempt fate. After all, there had been an Apollo 13 which lifted off at 1:13 back in April of 1970 and a lot went wrong. Now we were looking at "a date which will live in infamy" as Franklin D. Roosevelt had said back in 1941, the day after Pearl Harbor—December 7. If there were concerns, I never heard them expressed. On the contrary, since this was to be our last trip to the Moon for the foreseeable future, interest was higher than it had been for the last few missions.

The crew had picked the name America for the Command Module while the Lunar Module was named Challenger. Nearly 14 years later that name would become synonymous with one of the greatest disasters in NASA's history.

NASA had selected a region of the Moon known as Taurus-Littro because they expected to find rocks both older and younger than those brought back by previous missions. This was the third and last J-type mission. The letter "J" distinguished those three from all the others by virtue of the extended capabilities of certain hardware, a larger scientific payload capacity and the most obvious – the battery-powered LRV, Lunar Roving Vehicle.

The period after the CSM emerged from its Earth parking orbit and was boosted into a lunar trajectory was called "translunar coast time" and took a few days. It is during this part of the mission the Command and Service Module was docked with the Lunar Module, and the no-longer-needed S-IVB stage was then sent on a death plunge to lunar impact. This time it was noted and recorded by seismometers left on the Moon by Apollo 12, 14, 15 and 16.

Following successful lunar orbit insertion and separation of the LM from the CSM, the Lunar Module began its powered descent to the lunar surface and found Taurus-Littro at 2:55 p.m.on December 11.

They began the first of their three EVA's four hours later with offload of the Lunar Rover and the scientific equipment. The ALSEP package was deployed about 180 yards from the Challenger. Gene Cernan drove the Lunar Rover to that site where he drilled several deep core holes with the improved battery-powered drill. A far cry from the one Dave Scott had struggled with on Apollo 15. Cernan easily and often came up with core samples over 8 feet deep and that helped to make this last lunar mission the most productive of the entire Apollo program. Would they have wanted it any other way?

Over the course of those three EVA's, they took the Rover as far as 4 and a half miles from Challenger and picked up no less than 243 pounds of soil and rock samples.

Harrison Schmitt was the first true scientist to go to the Moon and, when, on the second EVA, they arrived at some obviously orange-colored soil, he was particularly excited because to his trained eyes this meant evidence of volcanic activity and probably fairly recent in the broad terms of life in the universe. Perhaps only a mere few hundred million years.

We know they were constantly picking up rocks of varying colors and textures, which indicated their possible chemical makeup. During their last trip out they went to the foot of a mountain where they saw, and vividly photographed, two large boulders that had obviously rolled down from the highest parts of the mountain. Large? They were huge. In one picture, one of them dwarfs the Lunar Rover parked beside it.

As they were winding up this last walk and ride on the Moon, Cernan attached and unveiled a plaque on Challenger's landing strut, or leg, which commemorated the completion of the Apollo missions. He then parked the little Rover at a spot where its TV camera could watch and transmit their takeoff. After that, the now routine business of trading off, the leaving behind of expendables in exchange for what they wanted to bring back from the Moon. The list of items sitting up there now and forever (?) includes any number of beautiful cameras, the bases of six Lunar Landers, three Lunar Rovers and all those seismometers, other scientific hardware, and at least half a dozen American flags wired to appear as though they are waving in that airless environment.

So far, no complaints about littering from anyone that we know of.

A very impressive departure as the LM lifted off from its base on December 14. TV audiences watched as the camera on the Lunar Rover, directed from Houston, followed their ascent stage out of sight. Then it turned and scanned the now-deserted lunar surface. They say you could almost hear the silence and absorb the Moon's quiet beauty.

Two hundred miles east of Pago Pago (that's pronounced Pango Pango, by the way) they splashed down at 2:24 p.m. on December 19 within a few miles of the recovery carrier USS Ticonderoga and its efficient group of recovery helicopters and frog teams. Ticonderoga had gotten a new skipper since Apollo 16, with Captain N.K. Green now on the bridge. HC-1 was the helicopter squadron for a third straight splashdown and its Recovery helicopter pilot was Commander E. E.

Dahill III who would go on to fly with that squadron on several Skylab missions. UDT-11, led by Lt. Jon Smart, provided the swimmers for the fourth time during the Apollo flights.

It had been a 12 day, 13 hour and 52 minute mission, and it closed a magnificent chapter in space history that may not be resumed for a long time.

Mercury, Gemini and Apollo, through the congressionally directed evolution of space hardware and systems into everyday life, have provided an incredible return on investment despite critics who have apparently never taken the time to investigate the real picture. A few more words on this subject on the final page of this book.

Before we forget, Apollo 17 – last on the Moon. Where is it today? Find it at the Johnson Space Center in Houston.

Added note: My exchange, sort of, with Gene Cernan. It was the summer of 1994 and I was asked to emcee an awards dinner at the Marriott Marquis Hotel on New York City's Times Square. The 25th anniversary of Apollo 11; and the Space Week Committee of Houston was giving framed awards to four gentlemen who, in their view, had contributed greatly to space reporting over the years. Walter Cronkite could not be there but Hugh Downs of ABC's 20-20 was there along with Jay Barbree of NBC, who has broadcast every manned mission since Alan Shepard and is still doing it to this day. Also on hand, Howard Benedict of the Associated Press/retired.

The Apollo 11 astronauts were not there but Gene Cernan, last man on the Moon was, as was Alan Shepard, first American in space. Both sat at our table. At one point in the evening I produced a small ceramic threaded bolt from the surface of the Apollo 11 Command Module and announced that the object in my hand had been to the Moon and back. Quietly, Cernan stood up, unzipped his smart black leather boots and tapped his ankle. "Look, ladies and gentlemen" he said, "this foot WALKED on the Moon." Oops. I knew when I had been topped. Gene did it with intense good humor, and we all had a good laugh. I still have that ceramic bolt and he's still walking around on that foot.

Ron Evans was a Navy flier and a Captain at the time of Apollo 17. Harrison Schmitt was a civilian.

Apollo	Dates	Astronauts	Highlights	Recovery Ship & Commanding Officer	Recovery Pilot	UDT Primary Unit
VII	1968 Oct. 11-12	W. Schirra, USN D. Eisele, USAF R. Cunningham, USMCR	1st manned Apollo in Earth orbit	Essex Capt. J.A. Harkins	Cdr. E. A. Skube, HS-5	21
VIII	1968 Dec. 21-27	F. Borman, USAF J. Lovell, USN W. Anders, USAF	1st manned Moon orbit	Yorktown Capt. J. G. Fifield	Cdr. D. S. Jones, HS-4	12
IX	1969 Mar. 3-13	J. McDivitt, USAF D. Scott, UASF R. Schweickart, Civ	1st lunar module flight, docked	Guadalcanal Capt. R. M. Sudduth	Cdr. G. M. Rankin, HS-3	22
X	1969 May 18-26	T. Stafford, USAF J. Young, USN E. Cernan, USN	1st Moon orbit by entire system	Princeton Capt. Carl Cruse	Cdr. C.B. Smiley, HS-4	11
XI	1969 July. 16-24	N. Armstrong, Civ B. Aldrin, USAF M. Collins, USAF	1st Moon landing and walk	Hornet Capt. C. J. Seiberlich	Cdr. D. S. Jones, HS-4	11
XII	1969 Nov. 14-24	R. Gordon, USN C. Conrad, USN A. Bean	2nd Moon walk	Hornet Capt. C. J. Seiberlich	Cdr. W. E. Aut, HS-4	13
XIII	1970 Apr. 11-17	J. Lovell, USN J. Swigert, Civ F. Haise, Civ	Malfunction Apr. 13; Mission aborted	Iwo Jima Capt. L. E. Kirkemo	Cdr. C. B. Smiley, HS-4	11
XIV	1971 Jan. 31 - Feb. 9	A. Shepard, USN S. Roosa, USAF E. Mitchell, USN	Mission to Fra Mauro	New Orleans Capt. R. E. Moore	Cdr. W. E. Walker, HS -6	11
XV	1971 July 26 - Aug. 7	D. Scott, USAF J. Irwin, USAF A. Worden, USAF	First use of Moon rover	Okinawa Capt. A. F. Huff	Cdr. S. A. Coakley, HC-1	11
XVI	1972 Apr. 16-27	J. Young, USN T. Mattingly, USN C. Duke, USAF	Record 71 hours on the Moon	Ticonderoga Capt. E. A. Boyd	Cdr. Arnie Fieser, HC-1	12
XVII	1972 Dec. 7-19	E. Cernan, USN R. Evans, USN H. Schmitt, Civ	275 pounds of Moon soil obtained	Ticonderoga Capt. N. K. Green	Cdr. E. E. Dahill III, HC-1	11

INTRODUCTION TO THE SKYLAB PROGRAM

To have a clearer understanding of the Skylab program, take a look at the Saturn Family. The big guy is in the middle. Saturn V, the monster that took all our astronauts to the Moon. On the right is

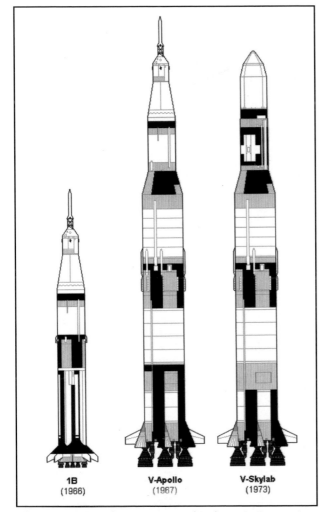

another full Saturn V "stack," as it was called, but this one had a new wrinkle. What had been the S-IVB section, the rocket's third stage, no longer carried fuel but had been converted into the Skylab workshop – a deluxe multi-story apartment in space that would have no rivals for room until the International Space Station came along years later. Because of its size and weight, it still required the power of the big Saturn V to attain Earth orbit. Remember, it was unmanned.

So NASA put the unmanned "apartment" into Earth orbit where it waited for the astronauts of Skylab 2, 3 and 4; and when the crews flew to their new space home, they got there atop the

The Saturn Family (Courtesy of Robert A. Braeunig Rocket & Space Technology)

Saturn 1B, shown at the left. After all, they were only using an Apollo Command Module as a mode of travel until they got to the orbiting workshop itself. They didn't need the awesome power of a full Saturn V any longer.

Now take a look at the S-IVB as it really looked inside a NASA workshop. That's the single J-2 engine at the bottom of the photo.

To complete the Skylab tour, here is an artist's concept drawing showing the two-story livable interior of what had been Saturn V's third stage and just beside it is another drawing

Saturn IV-B (Courtesy of NASA)

giving you a better idea of what the whole cluster looked like when the Service Module was attached to the docking adapter which, in turn, led the crews to the Orbital Workshop, S-IVB.

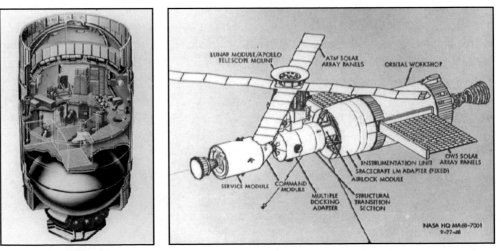

USS Ticonderoga was the recovery ship again.

Room to stretch out! Apollo was over. NASA decided, for a number of reasons, not the least of which was post-Vietnam war recession, to end the lunar landing missions with Apollo 17 when, originally, the space agency might have continued through Apollo 20. In other words, there was some unused hardware out there and with the Space Shuttle program still on the drawing boards, what would NASA do to implement a space development effort a lot sooner than the Shuttle, which did not fly until 1981?

The answer was Skylab and it was incredibly inventive. For an agency used to, but never fond of, the criticism which seemed to follow every mishap, the opportunity of using a few left-over Saturn rocket parts to create an early "space station" that was roomier than one the Soviets already had in space, should have silenced some of those critics.

Consider this. Take a complete Saturn V lunar mission package and pull it apart, placing each of its powerful segments a few feet apart from each other for the purposes of this illustration. You are now looking at the world's most powerful launch vehicle and its first three stages plus the CSM section at the very top. The CSM, the Command and Service Module, would be used in this new world; but the Service Module part previously used as a "garage" for LM, the lunar landing module, could now be diverted to other storage requirements that would allow much longer periods in Earth orbit.

As explained in the immediate preceding pages, what had always been a third stage rocket with its fuel tank became astronaut "housing." It became a big, friendly bear of a tube that must have seemed like a penthouse to any astronaut who had flown in the cramped confines of any of its predecessors. Now you had room for exercycles (ergometers), a rotating chair and countless other devices and medical experiments previously impossible to conduct, along with much longer missions in which to use and learn from all those experiments. It was a brilliant use of excess space hardware.

Skylab 1 was an unmanned flight that put this big new sky cabin into Earth orbit awaiting the arrival of the three crews to follow. It went up on May 14, 1973, and almost immediately had a problem. I often wonder if the gang back in Mission Control ever gets used to these roadblocks. It was on the way up when an external solar/meteoroid shield ripped off and the result was skyrocketing temperatures inside the station itself. Temperatures in which no humans could live and function. Consequently, before the first crew could even kick off their shoes and settle down in front of the TV (only kidding) they had a big repair job to do…. and they did it. After substantial repairs, the crew was able to deploy a parasol sunshade, which brought inside temperatures down to a very livable 75 degrees.

Skylab 2's astronauts took off for their hothouse in the sky on May 25 but because they were only going into Earth orbit and not all the way to the Moon, Saturn 1B, which had launched the early Apollo Earth orbital flights, was sufficient to get them up there.

How good was Skylab? Well, for openers, this mission would last 28 days, far and away this nation's best effort in that area. Loaded with all the medical monitoring equipment that it was, just imagine the sort of data NASA and its medical experts began collecting almost immediately on the really long-term effects of life in space. Now you had devices which tracked cardiovascular, musculoskeletal, hematological, vestibular, metabolic and endocrine systems of the body.

We should separate a few of those fancy terms for further explanation since a few flew right over our heads too. The metabolic analyzer measured oxygen consumption. The hematology side looked at the effects on the blood of weightlessness and time spent in space. Vestibular was the one that stopped us in our tracks. Actually it was a rotating chair which imposed accelerative forces on the vestibular organ of the inner ear. Welcome to Medicine 101.

There was lots of attention given to bone and calcium physiology, cardiovascular physiology, mineral balance, and the bioassay of body fluids (endocrinology), sleep monitoring, and time and motion studies.

Many experiments in Skylab did not directly involve the astronauts themselves, such as the survival and growth of bacterial spores in space; and this is one of the areas where high school students from across the country took part – from the ground, of course. NASA regularly invites our schools to come up with experiments that they would like to see carried out in space and, by the thousands, our young people have come up with projects, questions and problems which Skylab, the Shuttle and now the ISS have taken into space with them. For Skylab 2 there were five student experiments.

When it was over, Skylab had been up there for 404 orbits of the Earth, had logged 392 experiment hours, and conducted three EVAs that totaled six hours and 20 minutes.

At 9:49 a.m. on the morning of June 22, the Command Module for Skylab 2 splashed down in the Pacific some 800 miles southwest of San Diego and just about 5 miles from the recovery carrier USS Ticonderoga.

HC-1 (Helicopter Combat Support Squadron 1) also known as the "Fleet Angels" was there to pick up the astronauts. It was their fourth straight mission going back to Apollo 15 and their prime recovery pilot was Commander Arnie Feiser. The frogmen were members of UDT-11.

When we refer to returning Skylab astronauts, bear in mind they returned in Apollo Command Modules. The Lab itself remained out there for other crews to inhabit and years later would come back to Earth in a fireball.

The Skylab 2 CM – on view at the National Museum of Naval Aviation in Pensacola, Florida.

Paul Weitz came to the space program from a lengthy career as a Navy pilot but we could not find any source that would provide his rank at the time. Joe Kirwin was a Navy Captain and an MD and in charge of the extensive scientific experiments conducted on Skylab.

THE FLIGHT OF SKYLAB III WITH ASTRONAUTS ALAN BEAN, JACK LOUSMA AND OWEN GARRIOTT FROM JULY 28 TO SEPTEMBER 25, 1973

USS New Orleans was the recovery ship again.

Of mice and men. Look at those dates! From when to when? Fifty-nine days in Earth orbit. What do you do to keep busy? Well, for openers, you do a lot more of what has already been done, and then add in experiments predicated on situations or problems that appeared during a previous mission. Obviously, this mission moved to the head of the class in space duration for our astronauts.

This time they would bring along some bugs and some mice; hence the reference above. Both experiments were designed by high school students and would have produced some interesting information, but thanks to a power failure 30 minutes after the launch, both experiments folded. Fortunately, several other student experiments survived.

Now…back to the launch site, Launch Pad 39B. With the Skylab up there and waiting, the smaller Saturn 1B was now sufficient to get the modified Apollo Command and Service Module into Earth orbit, lift-off occurred at 7:11 a.m. on July 28.

Soon after they had docked with the Skylab, known as OWS for Orbital Workshop, all three had a bout with motion sickness but were able to get on with an ambitious schedule that not only resembled the schedule of experiments done by their predecessors, but added some valuable new ones.

On the fifth day of the mission, the crew discovered the apparent failure of two of the four thrusters on the CSM's reaction control system. If it had persisted, it could have meant "how do we get home?" But it also put in motion an ambitious effort at the Cape to ready a rescue mission, if needed. It was not. When all heads were put together, the decision was made to continue and the suspect thrusters worked just fine on their return to Earth.

On day number six, Jack Lousma and Owen Garriott took their first of three space walks (EVA's) lasting for six and a half hours. While they were outside, they deployed another sun shield to ensure continued habitable living conditions in the OWS and, as always, performed some external maintenance.

Meantime, the experiments into the effects of time and weightlessness in space continued. Heart, blood, bone density, hearing, and more. They looked into what was being called the "puffy face syndrome" first noted on Skylab 2. It had to do with the tendency for body fluids to head upwards in space.

Skylab was also reaching out with solar experiments and her cameras were working overtime. On one space walk they rolled out the AMU, the Astronaut Maneuvering Unit, which never got a chance to be used during Gemini 9 because of problems with the old Gemini space suit. Because of the AMU's appearance on Skylab, engineers were able to fine-tune the Manned Maneuvering Unit later used on a Shuttle flight in 1984.

The space walks came far apart. The first was on August 6 as noted, the second on August 24 during which film from solar telescopes was retrieved and replaced, and the third and last on September 22 when Alan Bean went out with Owen Garriott and they did more of the same.

There was a lot of television, and Earthbound viewers got some entertaining looks at the Skylab trio floating around in their workshop.

Time to go home. By the time they splashed down in the Pacific on September 25, they had orbited the Earth a tidy 858 times and had been out 59 days, 11 hours, 9 minutes and 4 seconds (but who's counting?). The amphibious assault carrier, USS New Orleans, which would finish up the "splashdown" part of our space program, was ready, waiting and able. Her skipper was Captain R.W. Carius, the helicopter squadron was once again HC-1 and its CO was Commander E.E. Dahill III. The frogmen were members of UDT-11 and their team leader was Lt. Jim Maxwell.

The Skylab 3's modified Apollo Command Module can be seen at NASA's Lewis Research Center in Cleveland, Ohio.

Jack Lousma came to the space program as a veteran Marine pilot with thousands of hours of flight time as did most of his fellow astronauts. Owen Garriott was a civilian and a scientist.

THE FLIGHT OF SKYLAB IV WITH ASTRONAUTS GERALD CARR, EDWARD GIBSON AND WILLIAM POGUE FROM NOVEMBER 16, 1973 TO FEBRUARY 8, 1974

USS New Orleans was the recovery ship again.

Astronauts have feelings too! From the moment Alan Shepard rocketed into sub-orbital space back in 1961, astronauts have spoken out, some rather loudly, about missions, equipment, training and their fellow astronauts and comrades in Mission Control. A few have been notably outspoken.

Now, as the final Skylab mission prepared to head into space for what would be America's longest manned event, over 84 days, the seeds had apparently been sown for dissension, and, ultimately, censure of three astronauts. I found this information both disturbing and fascinating since, as a ground-bound reporter back in New York City (NBC Radio Network) any reporting I did on Skylab 4 never mentioned any personal problems between the crew and Mission Control.

First, an effort to clarify one issue. On most websites you will find this mission described as Skylab 4, while on a few others it will be listed as Skylab 3, since it was the third manned mission of the four Skylabs. If you run across this oddity just keep in mind that most sources go with the "4".

Skylab 4 left the Launch Pad at 9 a.m. on the morning of November 16 with a whole lot of tasks lined up for the crew. The main purpose of this very lengthy mission was to do more of what previous Skylabs had done, and pile up the knowledge and experience of extended time in space and its effects on living, breathing human beings.

The reported "problems" started early in the mission when all three were in agreement in their griping about over- programming – what they quickly came to view as a seemingly endless string of tasks – and they made their feelings very clear.

While one report writer referred to blistering comments and grumpiness on the part of the astronauts and reaction to external authority as triggering ridicule, hostility and exasperation, another report writer preferred to call those comments "candid assessments" of faulty equipment. Sounds a lot like political "spin." Here again, when the crew suddenly decided to take a day off and step completely away from assignments, it was described as open revolt in one report, but pictured, in another report, as a measured effort to "put a hold" on activities pending a clarification of instructions and establishment of a suitable work pace. That could just be the most diplomatic description of a work stoppage ever put to print.

The good news is that relations between spacecraft and Earth people improved considerably, as did the crew's performance.

Four EVA's distinguished this final Skylab mission from all others with the expected new, higher numbers for hours spent outside the OWS. On all of these external events there was plenty of picture taking including some of the comet Kohoutek on at least two of those EVA's

They all involved maintenance, film changing and exposure experiments.

In addition to all the experiments on the crewmembers themselves described in previous mission reports, they had their usual student experiments on board; and they were motor sensory performance, plant growth and phototropism, the tendency of a plant to move toward or away from a light source such as the sun.

You would expect a few firsts in a mission of this length and you would be on target: longest space walk at just over seven hours, longest total space walk time of nearly 22 and a half hours, most scientific experiment hours at 1,563 and a whopping 34 and a half million miles zipping around the Earth – 1, 214 orbits.

And finally, despite their reported "grumpiness and weariness" early on, they came out of our longest effort in space in better physical condition than any previous Skylab trio except for loss of bone calcium, which was totally expected.

On station for another efficient recovery on splashdown day, February 8, 1974, was the USS New Orleans in the South Pacific but with a new Commanding Officer, Captain Neiger. HC-1 was called upon once again to provide the helicopters and Commander E. E. Dahill III continued as squadron CO while the prime recovery pilot was Commander W.D. Pocklington. This time is was UDT-12's turn to provide the frogmen and they were led by LTJG Pete Toennies.

Skylab 4 can now be found at the Smithsonian's National Air & Space Museum in Washington, D.C.

I mentioned crew censure in the beginning of this mission report. None of the three crewmembers of Skylab 4 ever flew in space again.

Gerald Carr had been a Marine pilot and was a Colonel at the time of this mission. Ed Gibson was the latest scientist to fly on Skylab and Bill Pogue came from the Air Force as a veteran test pilot with the rank of Colonel.

THE FLIGHT OF APOLLO-SOYUZ (ASTP) WITH ASTRONAUTS TOM STAFFORD, DEKE SLAYTON AND VANCE BRAND AND SOVIET COSMONAUTS ALEXEI LEONOV AND VALERY KUBASOV FROM JULY 15 TO 19, 1975

USS New Orleans was the recovery ship again.

Last splash!
While it wouldn't hold a candle to the Skylab missions for time spent in space, ASTP (Apollo-Soyuz Test Project) was a quantum leap in an area not normally dealt with or planned for in space missions – international cooperation and diplomacy.

Once again the tried and proven Apollo Command Module would take the ride, and, this time, dock with the Soviet Union's equally tested and reliable Soyuz spacecraft.

The U.S. chipped in with a new docking mechanism designed to leap over the physical differences, which normally would have prevented the space vehicles of the two nations from joining up. (See NASA diagram on next page)

This was an interesting crew all around. Tom Stafford, who flew on Gemini 9 and Apollo 10 as a Colonel, was now a Brigadier General in the Air Force. Deke Slayton, one of the original Mercury 7 astronauts, was on his inaugural flight into space and glad to be, having survived a medical problem – a heart condition that grounded him very early in the program. He had continued as the man whom all the others had to go through to get a seat in any manned missions. He waited a long time and said loudly and clearly shortly after lift-off, "Man, I tell you, this is worth waiting 16 years for."

On the Soviet side they had their own space pioneer in Alexei Leonov who, back in 1965, had become the first person, from any nation, to walk in space.

The Soviets started the ball rolling when they launched Soyuz from their Baikonur complex in Kazakhstan on July 15, 1975, and NASA sent up the Apollo CSM on the same day. Docking, as carried out by General Stafford was accurate to within one hundredth of a degree, the most accurate the Apollo system ever had.

The Americans were the first to open the hatch between the two spacecraft and they were immediately hit with what smelled like burning glue. It was escaping nitrogen tetroxide from a malfunctioning reaction control system. Bothersome at first, with coughing and difficulty breathing on the part of all the space men, but a quick dose of 100% oxygen cleared things up, and there were no further repercussions – even in post-flight physicals.

With that problem out of the way, it was time for the historic, first handshake in space between two old adversaries. Leonov and Stafford were first to "press the flesh" and the Soviets held up a sign "Welcome to Soyuz" as they invited our guys into their spacecraft. They exchanged souvenirs.

Despite the brevity of this mission which was, after all, primarily intended to prove the practicality of docking the space hardware of two competing countries, they were able to complete 28 experiments in the fields of Earth resources, gravity and atmosphere, astronomy, solar and life sciences and space processing. Once again, as on the last Skylab, an electric furnace was employed for experiments in solid materials processing. The five men got along well despite obvious difficulties with language. Tom Stafford had studied Russian and that certainly helped matters. As for Leonov, he would later say that the Americans were, in his words, "really nice guys" and they went to extremes to explain things in great detail. The Berlin Wall was still years from coming down but this was a beginning and a good one. When the Soviets put their Mir space station into orbit, American astronauts would be invited to visit and to observe and learn.

Later on, when plans got underway for the International Space Station, the U.S. and what is now Russia became the major contributors among a worldwide group of nations. NASA officials have always said the lessons learned from, and aboard, Mir proved invaluable in designing and building the ISS.

Another splashdown (the last one) took place in the Pacific about 300 miles west of Hawaii at around 1:15 p.m., July 24. For the last time the USS New Orleans was there to do the job. Also on hand to complete the recovery team, the helicopters of HS-6 and the frogmen of UDT-12*. The Soviets, of course, did their own thing and never did use the "splashdown" system. Many of their descents onto their own hard soil cost the

lives of some of their cosmonauts. Leonov and Kubasov, however, did get home safely.

Skylab decayed back into the atmosphere on August 2 and burned up. As for our part of the Apollo-Soyuz Test Project, the Command Module is now on display at the Kennedy Space Center Visitors Complex. You can't miss it.

The NASA community celebrated when Deke Slayton finally got to go into space. He was a veteran Air force pilot. Vance Brand was a Marine pilot with something close to 10,000 hours flying time.

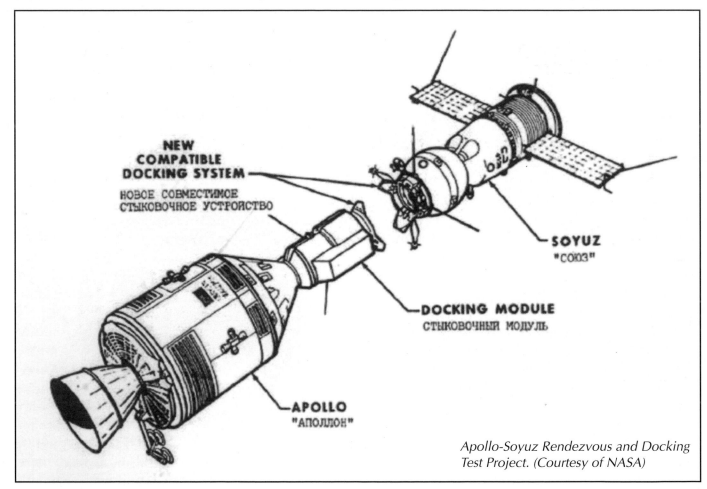

NEW COMPATIBLE DOCKING SYSTEM
НОВОЕ СОВМЕСТИМОЕ СТЫКОВОЧНОЕ УСТРОЙСТВО

SOYUZ "СОЮЗ"

DOCKING MODULE
СТЫКОВОЧНЫЙ МОДУЛЬ

APOLLO "АПОЛЛОН"

Apollo-Soyuz Rendezvous and Docking Test Project. (Courtesy of NASA)

---

*HS-6 and UDT-12 based on best available Internet data.

Apollo-Soyuz – artist's concept.
(Courtesy of NASA)

There you have it. Our retrospective look at, and tribute to, the men, women and machines that combined to provide their invaluable part of NASA's march to the Moon. We say men and women simply because neither NASA nor the Navy nor the DOD is a men-only club and there is no easy way of knowing just how many women toiled diligently behind the scenes while providing their assistance and expertise. Our hats are off to all who were involved and continue in that great work.

Along the way we have noted a few Spinoffs – namely some well-known migrations of NASA needs into the popular market place. The fact of the matter is, these "technology transfers" today number in the thousands and, by themselves, are more than ample justification for continuation of the nation's space program in both manned and un-manned modes. It is a message which has yet to be properly told. Sources far, far more expert than this writer have testified the ongoing return on investment in the space program from Day One ranges anywhere from 7 to 10 to one on the dollar. We can't think of a single government program that has achieved such incredible rate of return.

It is highly unlikely a day will go by without you and I having used, touched or benefited by some device, some technology or substance that is a direct pass-along from our space program. The ongoing outcry every time a satellite is lost or a launch has to be purposely terminated at its origin is invariably based on an honest enough concern for the almighty tax dollar but equally prompted by a deep-set ignorance of the truth.

The promise of the International Space Station, if the bean counters ever let it reach its full potential, is a cornucopia of riches such as no prior element of the space program has ever even approached. At this writing, those budget driven pseudo-experts are pecking away at the very foundation of this incredible resource. Laboratories that, if left unhindered, will give us breakthroughs in medical, industrial, agricultural, meteorological, and other fields.

Before you join the chorus of critics with the old critique about how many hungry mouths might be fed with NASA's dollars, remember the above statement about a return of 7 to 10 to one on the dollar. Additionally, NASA's budget never exceeded one percent of the Federal budget.

Keep these numbers in mind and the next time the ISS cruises over your neighborhood, give the crew a thumbs up. They all know that just getting them up there and then bringing them back is fraught with incredible risk, and we know the price too many of them have already paid. R&D never has been, nor ever will be, undertaken without those risks.

Jim Blair

---

* Could there ever be another splashdown? It's possible now that President George W. Bush, on January 14, 2004, has spoken the words, "Let us continue the journey." Of course that aim could be disregarded, eliminated completely, or enthusiastically embraced by future administrations. It did receive fairly widespread support at the time those words were spoken, and they included the prospects of building a station on the Moon itself and using it as a base for deeper space exploration. The President also said we must complete the International Space Station, and that alone would mandate bringing the currently-existing fleet of three shuttles back up to safe flight status. No amount of economic and budget difficulties here on Earth should deflect this nation from that more immediate goal.

ACKNOWLEDGMENTS

We got off to a slow start on this book sometime back in 2001 and for a long time I even entertained the thought of completing the task with just my trusty computer. NOT! That might have been true for the latter portions of this book, where the condensed histories of each manned mission and the recovery ships involved, appear in consecutive mission order.

All that changed drastically when we dug into our more personal involvement in the downrange experience and it led to a steady drumbeat of old acquaintances via e-mail plus a growing roster of new friends by the same route. They became invaluable resources, never stingy with their time or their recollections, and I sincerely hope their order of appearance in the following listing is not interpreted as any degree of importance one way or the other. It would be impossible to categorize.

Nonetheless, it is very difficult to get a book out without a publisher; so Jack Sauter, keeper of the flame for the USS Lake Champlain website, goes to the top of our list and got there the moment he put me in touch with Turner Publishing. We had an over-the-phone tacit agreement to publish just a day after Sauter's phone call.

Bob Fish is curator of the USS Hornet Museum in Alameda, California. I was e-mail directed to Bob by historian Michael Esslinger who had played a key role in saving the Hornet from the scrap heap. From that moment on, Bob Fish rarely missed a day without firing off some vital new information and was still at it mere hours before we headed for the Turner offices in Kentucky with our completed manuscript. He did not stop there however. We were considering using a NASA photo of Apollo 16 at the instant of splashdown for the cover. Bob introduced me to world-renowned aviation artist Stan Stokes and his magnificent work entitled "All Navy Team." As you can see, this rare piece of "recovery art" captures the essence of what this book is about. It was everything we hoped the book would turn out to be. A spacecraft returning from the Moon, the prime recovery carrier USS Hornet and the celebrated helicopter #66 of HS-4. NASA and the Navy.

But long before Jack Sauter, Michael Esslinger and Bob Fish entered the picture, there were a few former frogmen, UDT's, starting their own steady stream of recall. Denny Bowman is now a Texan, Chris Bent and Bob Petersen are fellow Floridians, and my debt to all of them is wide and deep.

Dr. John Stonesifer of NASA, and the space agency's top man on so many carriers, also climbed aboard with page after page of priceless recollections. Between Jennifer Troxell at NASA Hq in Washington, D.C., and Shelly Kelly (neat name) over at the University of Houston, Clear Lake, we were presented with a wonderfully complete listing of every ship which took part in every mission. They even led us to all the ships involved in unmanned recoveries and we decided, why not, this IS a tribute to the Navy's role in the space program.

Captain Gordon Hartley, skipper of the Wasp for several of its Gemini recoveries, and a friend since the day I boarded his ship in Boston in 1966, always responded to my e-mailed inquiries. Carl Seiberlich was Captain Seiberlich when he took the Hornet out to bring home the astronauts of Apollo 11 and 12 in 1969. When we spoke over the phone in the fall of 2003, he was an Admiral in semi-retirement and still glad to talk about splashdowns.

Sikorsky Aircraft's Bill Tuttle, Director of Humanitarian Communications, chipped in with comments and corrections on my description of the helicopters that dominated the recovery program. He further improved our efforts when he offered two wonderful photos of the company's founder with some rather well known astronauts.

Mike Smolinski is responsible for just about all the patches you will find in these pages for all the ships, helicopter squadrons and UDT units. Important contributions came from NASA's Kay Grinter, Elaine Liston and Barbara Greene at the Cape and Annette Amerman at the Marine History offices in Washington. Also Dennis Ranalla of UDT-13 and Terry Muehlenbach of UDT-11.

The beat goes on. E-mails, photographs and charts day after day, week after week. Questions answered, mysteries solved by the likes of Bruce Sharpe, Steve Roberts and David Goodale on Naval matters, Milt Putnam – a Navy combat cameraman with HS-4 on the recoveries of Apollo 8, 10 and 11 and was one of only two photographers deployed over the splashdown scene in the "Photo" helo. Charles G. Farrant at the New Mexico Museum of Space History, Allan Needell at the Smithsonian Air and Space Museum, Mark Roberts at DestroyersOnline, Marine helicopter pilot Jim Lewis and Ron Phipps,

both of Marine squadron HMR (L)-262, Steve Alford at NASA, Gary Dobias (Hawaii), former frogman Gene Warta, Gary Kitmacher - NASA Houston, and John Hargenrader, NASA Hq.

Well before the final pages had fallen into place, my wife and bookkeeper, Patricia, stepped in and judiciously exorcised my punctuation demons, omissions and assorted missteps and offered frequent suggestions which were on the money more often than not. Yes, I am one of those men who won't stop to ask for directions. She has also kept me out of Chapter 11 for going on to 40 years.

The thought prevails, as we close these pages, that we have left someone out, perhaps more than a few. There were so many. Perhaps if my record-keeping habits were better than they are there wouldn't be any doubts. I've got my fingers crossed.

INDEX